URANIUM DEPOSITS:
Origin, Evolution, and Present Characteristics

An analysis of the present uranium deposits in terms of the geometrical, mechanical, thermal, and chemical aspects of the Earth's behavior during the past 4.6 billion years.

J. H. TATSCH

TATSCH ASSOCIATES
Sudbury, Massachusetts 01776
1976

Printed in the United States of America, February, 1976. Copies are available from Tatsch Associates or from book dealers in all parts of the world.

International Standard Book Number: 0-912890-11-8.

Library of Congress Catalog Card Number: 75-9304.

Preface

The increasing demand for energy and the decreasing resources of organic fuels make uranium an important consideration for future energy requirements. The importance of uranium in this connection was emphasized recently by the sale of almost half a million kilograms of uranium for just under $100 per kilogram, or $40 per pound.

The worldwide resources of uranium amount to roughly two million kilotons of yellowcake (U_3O_8), enough to supply the world's anticipated needs for roughly ten years. Requirements beyond that are so great that new exploration and ore-finding techniques will have to be developed.

It has been known for some time that many uranium deposits appear to be associated with plate boundaries. But what about metallogenesis before the first plates existed? Some of the Earth's uranium deposits appear to predate the earliest plates. To avoid this dichotomy, this book uses a new hypothesis for the origin, evolution, and present characteristics of the Earth's internal behavior, dating back to the "earliest Archean," roughly 4.6 billion years ago, hundreds of millions of years before the first plates are thought to have formed. This new hypothesis, derivable from a "tectonospheric Earth model," is described in the book, *The Earth's Tectonosphere,* and used as the basis for the books, *Mineral Deposits* and *Petroleum Deposits,* by the same author.

Even a precursory analysis of the Earth's uranium deposits reveals that they are not homogeneously distributed. A closer scrutiny suggests that the distribution patterns of these deposits are relatable to certain geometrical, mechanical, thermal, and chemical heterogeneities that have developed within the upper 1000 kilometers of the Earth during the 4.6 billion years that the Earth is believed to have been in existence.

The basic hypothesis of this book performs, essentially, a two-fold function in connection with these heterogeneities: (1) it provides a long-lived, deep-seated, global mechanism for their origin, evolution, and present characteristics; and (2) it provides a means of predicting the locations and characteristics of other heterogeneities that may be associated with hidden uranium deposits. The ability of the Tectonospheric Earth Model to perform these two functions provides a means of better understanding the origin, evolution, and present characteristics of uranium deposits. This, in turn, forms the basis for using the Tectonospheric Earth Model as a supplementary tool in the exploration for uranium deposits.

Table of Contents

Chapter 1

URANIUM DEPOSITS: A GLOBAL SURVEY

Uranium is a silvery white metal consisting of three semistable isotopes, U-234, U-235, and U-234. Its importance as an energy source derives from the release of heat associated with the fission of U-235, which comprises only about 0.7% of natural uranium. Most of the remaining 99.3% is U-238, because U-234 comprises only about 0.005% of the total. The abundant U-238 is not readily fissionable, but it can be converted, by bombardment, to the fissionable Pu-239.

Used prior to 1942 primarily for glass coloring, uranium became important with the advent of controlled nuclear fission during that year. This added two important uses: (1) as an explosive and (2) as a source of heat. With the improvement of nuclear reactors for generating electricity, uranium promises to become one of the world's important energy sources.

The earliest discovered uranium deposits were in Czechoslovakia (1727) and in other parts of Europe. The first extra-European sources of uranium were the uranium-vanadium sandstone deposits in western Colorado and eastern Utah, U. S. A. In 1923, the large, rich Shinkolobwe vein deposit was discovered in the Belgian Congo. Seven years later, the Eldorado vein deposit was discovered at Port Radium, N. W. T., Canada. Increasing requirements for uranium during the past 30 years have spurred exploration and discovery of deposits in various parts of the world.

But not all provinces have responded equally well to increased exploratory efforts. This is due primarily to the variation of uranium within different rock types. Consequently, over 95% of the reported uranium reserves occur in only six metallogenic provinces (Brinck, 1974):

Colorado Plateau and Wyoming	30%
Witwatersrand Basin	23%

Canadian Shield	21%
Northern Territory, Australia	8%
Niger, Gabon, Central African Republic	8%
European Hercynian belts	6%

Uranium Content by Rock Type.

Considered on a worldwide basis, the average uranium content of some common rock types and sediments may be listed:

Dunite, peridotite, pyroxenite, and other ultramafics	0.03 ppm
Basalt, gabbro, and other mafics	0.05 ppm
Andesite, diorite, and other intermediates	1.8 ppm
Clay, shale, and other sedimentaries	3.2 ppm
Granite, granodiorite, and other felsics	3.5 ppm

Other things being equal, the highest concentrations of uranium appear to be within the youngest members of each of these rock types. Based on an average of one part mafic and two parts felsic composition, the Earth's crust averages roughly 2.35 ppm (parts per million) uranium.

Types of Uranium Deposits.

Although uranium occurs widely on all continents, the highly-concentrated minable deposits are rather rare. Minable concentrations range from 400 to 2500 times the crustal abundance of 2.35 ppm. The favorable rocks and geologic environments are limited to less than ten major types. These include the following:

a. Uranium deposits in quartz-pebble conglomerates.
b. Uraniferous vein deposits.
c. Peneconcordant uraniferous deposits.
d. Other epigenetic uraniferous deposits.
e. Uraniferous igneous rocks.
f. Uraniferous phosphatic rocks.

g. Uraniferous marine black shales.

No one knows exactly how these deposits have originated, have evolved, and have become emplaced into their present locations. But it is well to take a brief look at what is known and what has been postulated in connection with these seven major types of uranium deposits.

Uranium Deposits in Quartz-Pebble Conglomerates.

Uranium deposits of this type are very large and extensive. The largest known deposits include those in Canada (Blind River - Elliot Lake), Africa (Witwatersrand), and Brazil (Belo Horizonte).

These uraniferous Precambrian quartz-pebble conglomerates were deposited under deltaic or fluvial conditions within shallow basins located in cratonic or marginal cratonic areas more than 2.3 b. y. (billion years) ago. Most observers feel that, during nearly oxygen-free reducing atmospheres of early Precambrian times, these deposits evolved when rounded, polished detrital grains of uraninite and pyrite accumulated with the more typical placer minerals.

According to the Tectonospheric Earth Model, these grains of uraninite and pyrite were vestiges of the "original" Archean uranium that had been "reworked" several times during episodes of seismo-tectonomagmatic-belt activity during the Archean and early Proterozoic (Tatsch, 1973a). Some of the placers of uraninite and pyrite grains were later modified by subsequent seismotectonomagmatic-belt episodes. However, both these minerals are generally coextensive with conglomerate beds or coarse parts of conglomerates.

In the Blind River - Elliot Lake district of Canada, unaninite-brannerite deposits are two to five meters thick and as much as 2 km across. These deposits contain more than 5 million tons of ore, averaging 0.12 to 0.16 % U_3O_8. Gold ranges from 0.02 to 0.03 ounces per ton.

In the Witwatersrand area of South Africa, the deposits are

larger and more extensive than are those in the Blind River - Elliot Lake area of Canada. However, they are of a lower grade, averaging about 0.03 to 0.07% U_3O_8. The Witwatersrand uranium is produced mainly as a by-product of gold (See, e. g., Tatsch, 1975f).

According to the Tectonospheric Earth Model, similar uranium deposits in quartz-pebble conglomerates should exist also within the vestiges of Archean and early Proterozoic seismotectonomagmatic-belt activity of each of the Earth's continents (Tatsch, 1973a).

Uraniferous Vein Deposits.

Uraniferous veins are found in a variety of geologic settings representing vestiges of seismotectonomagmatic-belt activity ranging from the Precambrian to the Tertiary. Most of these veins are fissure fillings in faults, joints, and fracture zones. The ore occurs as tabular bodies, irregular stockworks, pipelike masses at fracture intersections, and within mineralized gouge and breccia. The primary types of alteration associated with the veins include argillic, chloritic, hematitic, and sericitic systems. Veins range from a few cm to about a meter in width and extend as much as 100 m downdip and along strike. The resulting vein systems, comprising many veins, may extend to depths of almost 2 km (e. g., Canada).

The mined ore averages from 0.1 to 1.0% U_3O_8. The largest systems associated with base-metal sulfides include those ar Beaverlodge (Saskatchewan), Joachymov (Czechoslovakia), Marysville (Utah), Port Radium (N. W. T., Canada), and Shinkolobwe (Zaire).

The Bancroft (Ontario) uraninite-thorite veins and associated pods cut and replace the unzoned granite pegmatite bodies. These deposits average about 0.1% U_3O_8 and from 0.025 to 0.2% ThO_2.

Most observers feel that the uraniferous veins were deposited from hydrothermal solutions differentiated from magmas.

Peneconcordant Uraniferous Deposits.

Peneconcordant uraniferous deposits are widely distributed within the vestiges of Paleozoic-to-Tertiary seismotectonomagmatic-belt activity in various parts of the world. This includes those of the western U. S. A. In most of these, the uranium occurs in sandstone lenses interbedded with mudstones. The stratal systems appear to have formed under fluvial, lacustrine, and near-shore marine conditions in cratonic or marginal-cratonic environments. Most of the sandstone beds are quartzose sediments derived from older sedimentary rocks. Some arkosic sediments were derived from granitic rocks. Volcanic ash appears in some of the uraniferous sandstone beds and in the associated mudstones. Most of the beds contain organic residues or carbonized plant matter. Most of the uraniferous sediments are localized within one (or a few) stratigraphic units.

The primary uraniferous ores are uraninite and coffinite. Some deposits contain also carnotite and other secondary minerals. Most of the uranium ores occur within the sandstone pore spaces; but in some the ores replace the sand grains or the carbonized plant matter.

The mined ore ranges from about 0.15 to 0.30% U_3O_8. The size of the uraniferous bodies ranges from two-ton masses to those of more than 10 million tons. The forms of the ore bodies are usually either tabular or roll bodies. The tabular bodies are nearly concordant with the host sedimentary structures. The crescent-shaped roll bodies are elongate in plan and discordant to the bedding in cross-section. The tabular bodies form discrete masses that tend to cluster in favorable areas measuring a few km across and enclosed in reduced sandstone. The roll deposits occur intermittently along the crescent-shaped interfaces wedged between oxidized sandstones on the concave face and reduced sandstone on the convex face of the crescent. These interfaces may extend for several km.

Large tabular deposits include those at San Juan Basin (New

Mexico) and Uravan (Colorado). Typical roll deposits include those at Shirley Basin (Wyoming) and Coastal Plains (Texas).

No one knows exactly how the uraniferous peneconcordant deposits have originated, have evolved, and have become emplaced into their present tabular and roll environments (See, e. g., Finch, 1967). Most observers feel that the uranium was leached (1) from volcanic glass associated within or above the host rock, or (2) from granitic terraces exposed along the margins of sedimentary basins. Then the uranium was transported by ground water in the hexavalent (U^{6+}) state until precipitated under reducing conditions. The possible reducing agents include (1) carbonaceous matter, (2) H_2S gas, and (3) sulfite derived from the oxidation of preore pyrite.

Under the Tectonospheric Earth Model concept, all these conditions would be expected within vestiges of seismotectonomagmatic-belt activity (Tatsch, 1973a). The tabular uraniferous bodies would be wholly enclosed in reduced rock, sometimes surrounded by oxidized rock not in contact with the ore. The roll-type uraniferous deposits would form at a dynamic interface between oxidizing and reducing conditions, extending downdip.

Other Epigenetic Uraniferous Deposits.

Not all epigenetic uraniferous deposits can be categorized within one of the above three types. It is well to list a few from this miscellaneous epigenetic group:

a. Gunnar (Saskatchewan) is a multi-million-ton irregular pipelike body of uraninite (UO_2) and uranophane ($Ca(UO_2)_2Si_2O_7 \cdot 6H_2O$).

b. Nabarlek (Australia) is a fairly new deposit of epigenetic uranium in older Precambrian rocks. Two similar bodies are Jim Jim and Ranger, both also in the same region of the Northern Territory as is Nabarlek.

c. Midnite (Washington) is in Precambrian sedimentary rocks along contact with granitic vestiges of Cretaceous seismotectonomagmatic-belt activity. Some ore bodies measure 60 m wide, 200 m long, and 50 m thick. The average mined grade is about 0.23% U_3O_8. The uraniferous deposits appear to have formed shortly after granitic intrusions associated with Cretaceous seismotectonomagmatic-belt activity. Apparently, hydrothermal solutions were directed along favorable zones adjacent to the intrusive contact (See, e. g., Bacroft and Weis, 1963).

d. Ross Adams (Alaska) comprises numerous uraninite and throium-bearing veinlets cut by perkaline granite. Within these, there is an abundance of urano-thorite and uranoan thorianite formed as late-crystalizing accessory minerals (See, e. g., MacKevett, 1963). Ross Adams has yielded over 75,000 tons of ore, averaging more than 0.60% U_3O_8.

e. Rabbit Lake (Saskatchewan) is a subhorizontal, irregular gular, tabular body of uraniferous metamorphic rock.

Other mis-fit epigenetic uraniferous deposits are found in other parts of the world. Although these bodies do not fit into any of the simpler conventional categories, all of them are identifiable with vestiges of seismotectonomagmatic-belt activity, ranging from the Archean to the Tertiary (Tatsch, 1973a).

Uraniferous Igneous Rocks.

Some pegmatites and other igneous rocks contain uranium as an indigenous or genetic constituent. Some of the pegmatites and other hosts are of the zoned type. Other rock types include alaskites (e. g., Rössling, S. W. Africa); biotites (e. g., Conway, New Hampshire; pyrochlore-bearing alkalic rocks (Araxa, Brazil; and Ontario, Canada); riebeckite granite (Nigeria); and sodalite foyaite and nepheline syenite (Julianehaab, Greenland).

Some of these deposits contain low concentrations of uranium dispersed over wide areas. Conway, New Hampshire, falls into this category. There, the biotite phase of the Triassic-Jurassic granite averages 0.0015% U_3O_8 over an area of roughly 800 km^2. Other deposits of this type contain richer but smaller deposits. According to the Tectonospheric Earth Model concept, these deposits are vestiges of part of the Earth's "original" Archean uranium that have been "reworked" by Precambrian and Phanerozoic seismotectonomagmatic-belt activity. In the case of the Conway deposit, the most recent seismotectonomagmatic-belt episode occurred during the mid Mesozoic. The Conway deposit has probably been reworked more vigorously than have the more-concentrated less-dispersed deposits, such as those in Araxa, Ontario, Nigeria, and Julianehaab.

Uraniferous Phosphatic Rocks.

Most phosphatic rocks are uraniferous and show a positive correlation between the uranium and phosphate contents. Thus, marine phosphorite, the dominant source of phosphate, provides a very large resourec of uranium (See, e. g., Brobst and Pratt, 1973).

Uraniferous marine phosphorite deposits, usually 2 to 4 m thick, underlie hundreds of sq km of the Earth's surface in various parts of the world. These average from 0.007 to 0.07% U_3O_8. Some typical deposits include the phosphorite formation that underlies about 400,000 km^2 of Idaho, Montana, Utah, and Wyoming (See, e. g., McKelvey and Carswell, 1956). The uranium content ranges from 0.001 to 0.075% U_3O_8. The richest beds are those thicker than 1 m and containing 30% P_2O_5; these average 0.012 to 0.02% U_3O_8.

Phosphorite in the Pliocene Bone Valley Formation, Florida, is about 3 m thick over an area of roughly 500 km^2. It averages 0.012% to 0.024% U_3O_8.and 20% to 30% P_2O_5 (Altschuler et al., 1956).

Large portions of the marine phosphorite along the Mediterranean from Morocco to Israel contain upwards of 0.01% U_3O_8 (Davidson and Atkin, 1953).

Some of the depression-filling phosphorites in central Africa are rich in uranium (Mabile, 1968). Farther west, in Senegal and

Nigeria, the aluminum phosphate deposits are uraniferous.

Deposits of phosphorite near Recife, Brazil, contain 0.02% U_3O_8. These deposits resemble those at Bone Valley.

The exact origin of the uranium within phosphorites is not known. But it is known that, within phosphatic rocks, the facies that are associated with vestiges of seismotectonomagmatic-belt activity are more uraniferous than are other facies. Thus, the geosynclinal facies within phosphatic rocks are more uraniferous than are platform facies. Uranium in some marine phosphorite deposits was probably deposited from sea water during sedimentation. In other marine phophorite deposits, the uranium was deposited later by downward-percolating ground water.

Uraniferous Marine Black Shales.

Many marine black shales rich in organic material are uraniferous. The uranium appears to have been deposited under anaerobic conditions with the organic factor of the shale within shallow-water epicontinental seas. Most of these deposits formed after seismotectonomagmatic-belt activity during the early and middle Paleozoic. These include Devonian deposits in Tennessee, Alabama, and Kentucky and Cambro-Ordivician deposits in Sweden and the U. S. S. R.

Chapter 2

THE TECTONOSPHERIC EARTH MODEL: A NEW CONCEPT

One of the theses of this book is that the Tectonospheric Earth Model (Tatsch, 1972a) may be used as a supplementary tool in the exploration for uranium deposits. The reader not familiar with this model should consult the cited text for the geometrical, mechanical, thermal, and chemical details regarding the model. For the reader already familiar with the model, this chapter is intended to serve as a review.

The Basic 5400-km Primordial Earth.

When a dual primeval planet hypothesis is applied to an analysis of the early solar system, one of the consequences is a primordial Earth and a co-orbital companion, Earth Prime. Each of these modified Bullen-type bodies had a radius of about 5400 km and was fractured into octants early in the history of the solar system. The 5400-km primordial Earth became the "subtectonosphere" of the present Earth. The Earth's tectonosphere (the upper 1000 km in this model) is an accretion onto the primordial Earth of the remnant substance of five of the octants of Earth Prime, which had separated into octantal parts early in the history of the solar system. The early Earth may be represented schematically as shown in Fig. 2-1.

The center, C, represents a solid core, with an undefined radius assumed to lie somewhere within the range 0 to several hundred km. The hatched spherical shell, MC, represents a molten core. The 8 octants of the primordial mantle surround the molten core and are marked S. The preferential flow of heat, and possibly of volatiles, outward from the fracture belts, F, causes igneous activity at I.

The primordial Earth has a volume such that, when 5/8 of an

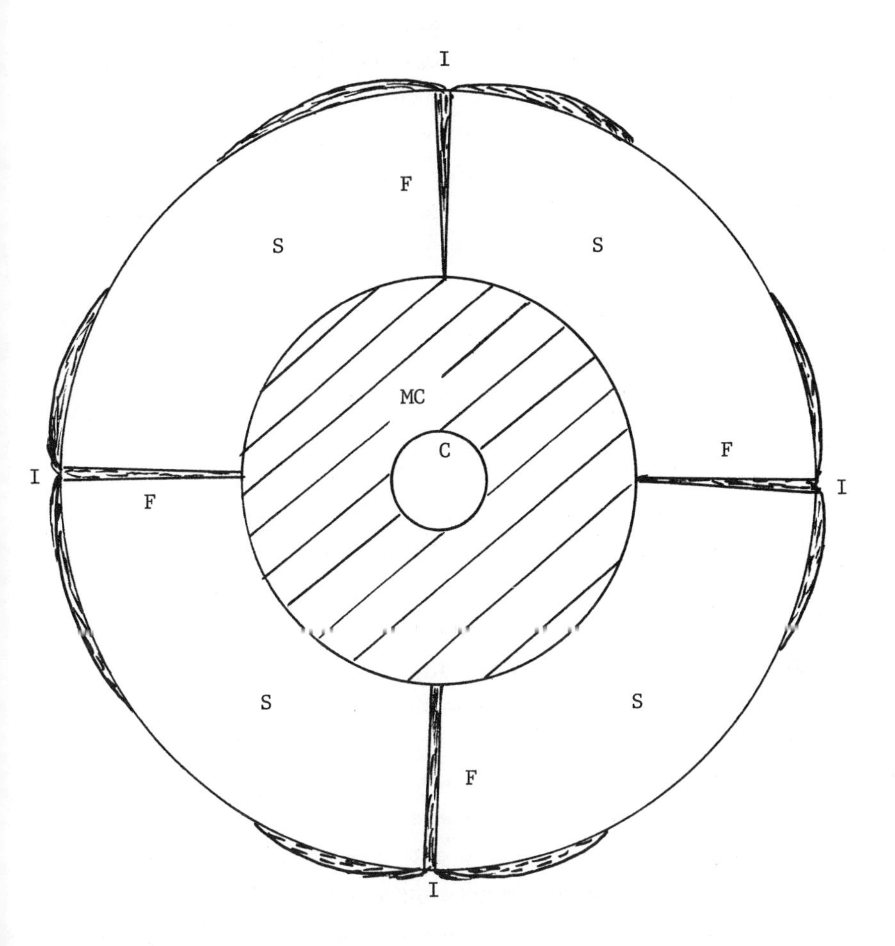

Figure 2-1. The 5400-km primordial Earth which, according to the Tectonospheric Earth Model, has been cut by three mutually orthogonal planes passing through the center. Schematic; not to scale. See text for details.

equivalent volume is added, the resulting body has a radius of about 6371 km, the radius of the actual Earth. This gives the primordial Earth a radius of about 5400 km, thus:

$1\ 5/8\ V' = V$, where V' and V are the volumes of the primordial Earth and of the actual Earth, respectively. Then, since

$V = 4/3\ \pi R^3$, we may write

$1.625\ (R')^3 = R^3$, from which

$R' = (1.625)^{-1/3}\ R$

$R' = (1.625)^{-1/3}\ (6371\ \text{km}) \doteq 5400\ \text{km}.$

This is the radius of the primordial Earth, as well as of Earth Prime and of the subtectonosphere used in the model.

The thickness of the tectonosphere is about 6371 km minus 5400, equals 971, or about 1000 km, roughly. Although the values, 5400 km and 1000 km, are used in this model for the radius of the subtectonosphere and for the thickness of the tectonosphere, respectively, it should be pointed out that these values are approximations which depend somewhat upon the particular assumptions made in connection with the assignment of initial values to the variable parameters of the basic tectonospheric Earth model derived from the dual primeval planet hypothesis.

Since the primordial Earth is fractured into octants by 3 mutually-orthogonal central planes, its surface is encircled by 3 mutually-orthogonal great-circular fracture belts, each having a length of 2 pi times 5400 km = 34,000 km, roughly, or about 3 x 34,000 km = 102,000 km total for the three belts.

The Geometrico-Mechanical Behavior of the Basic 5400-km Primordial Earth.

In order to analyze the geometrico-mechanical behavior of the basic 5400-km primordial Earth, it will be assumed that the basic behavior of the primordial Earth is such that it tends to equilibrate itself to a state of minimum energy. The driving mechanism of the system may be expressed as a function of the disequilibration energy inherent in its initial state. In simplest terms, it consists essentially of the resultant of two factors, each of which has been varying in a predictable manner during the past 4.6 b. y.: (1) the potential energy inherent in the geogenetically disequilibrated shape of the basic 5400-km primordial Earth; and (2) the selectively-channeled energy from the preferential flow of heat, and possibly of volatiles, outward from the three mutually-orthogonal belts which are "active" along the tensile portions of the surficial traces of the 3 planes separating the solid portions of the primordial Earth into 8 octantal subtectonospheric blocks.

When the driving mechanism of the basic model is applied to the primordial Earth, the behavior will depend upon certain geometrico-mechanical constraints inherent in the model. The simpler of these constraints may be summarized in terms of several types of equilibration, or tendency to assume a status of minimum energy:

Rotary Equilibration.

Fig. 2-2 shows schematically the expected basic behavior of the original 5400-km primordial Earth in response to the fundamental driving mechanism of the model from the standpoint of rotary equilibration. Basically, there are three degrees of internal rotational freedom, constrained by the configuration of the model to act along the three mutually-orthogonal planes along which the octantal fracturing occurred.

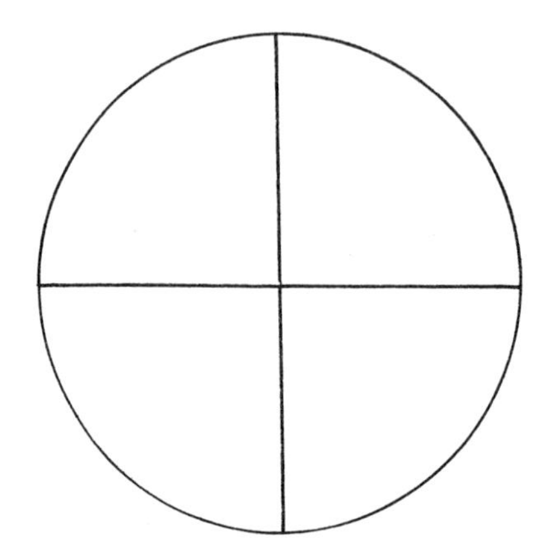

A. No rotation.

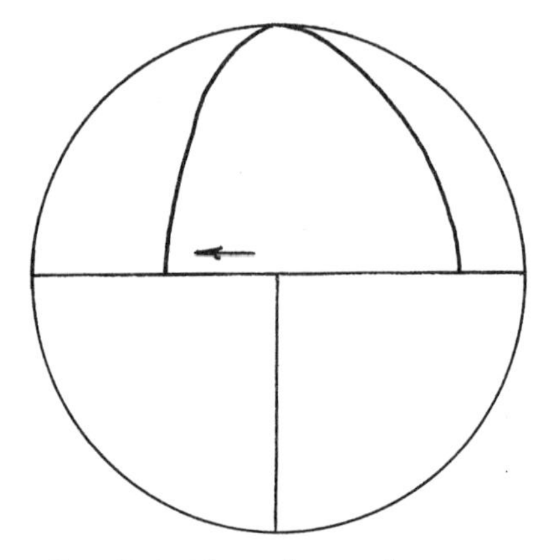

B. Rotation along the geophysical equatorial plane.

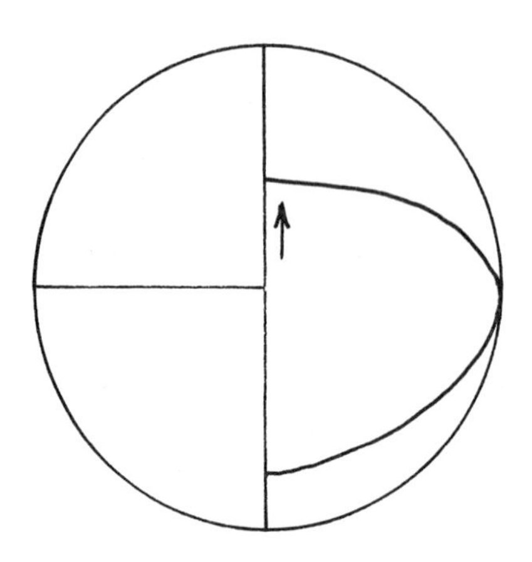

C. Rotation along the geophysical prime meridian plane.

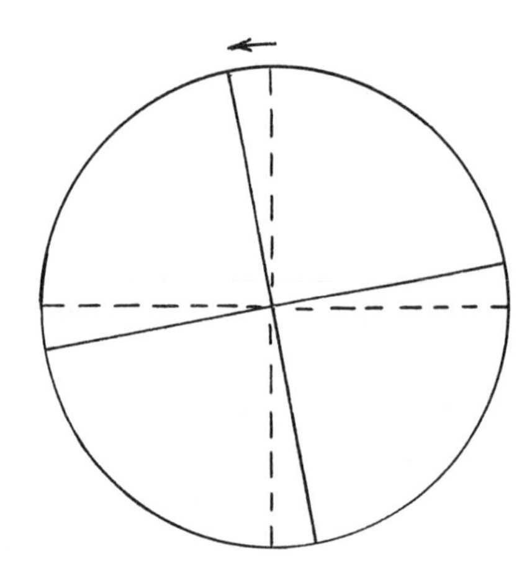

D. Rotation along the geophysical orthogonal meridian plane.

Figure 2-2. The expected basic behavior of the original 5400-km primordial Earth in response to the fundamental driving mechanism of the Tectonospheric Earth Model from the standpoint of rotary equilibration. Schematic; not to scale. See text for details.

It is noticed that, once the rotational motion has begun in any one of the three mutually-orthogonal fracture planes, the gross rotational motion in the other two planes is restricted until a net rotation of n pi/2 shall have transpired in the first plane, where n = 0, 1, 2, 3, - - -.

Radial Translatory Equilibration. Fig. 2-3 shows schematically the expected basic behavior of the original 5400-km primordial Earth in response to the fundamental driving mechanism of the model from the standpoint of radial translatory equilibration. The spherical octants are free to move radially with respect to the center of the model. From the principle of hydrostatic adjustment applied on a global basis, about half of the octants are subsided, at any one time, while the others are elevated with respect to the basic 5400-km radius. The magnitude of the radial motion visualized here is of the order of cm/yr, with maximum amplitudes of the order of ten km.

Transverse Translatory Equilibration. Fig. 2-4 shows schematically the expected basic behavior of the original 5400-km primordial Earth in response to the fundamental driving mechanism of the model from the standpoint of transverse translatory equilibration. Within the constraints of the model, some of the octants (normally two or four) may move transversally along the three mutually-orthogonal fracture-planes. The magnitude of this transverse motion is of the order of cm/yr, with maximum amplitude of the order of km.

Individual Octantal Motion. Fig. 2-5 shows schematically the expected basic independent behavior of the individual octants of the original 5400-km primordial Earth. Within the constraints of the model as a whole, each of the spherical octants

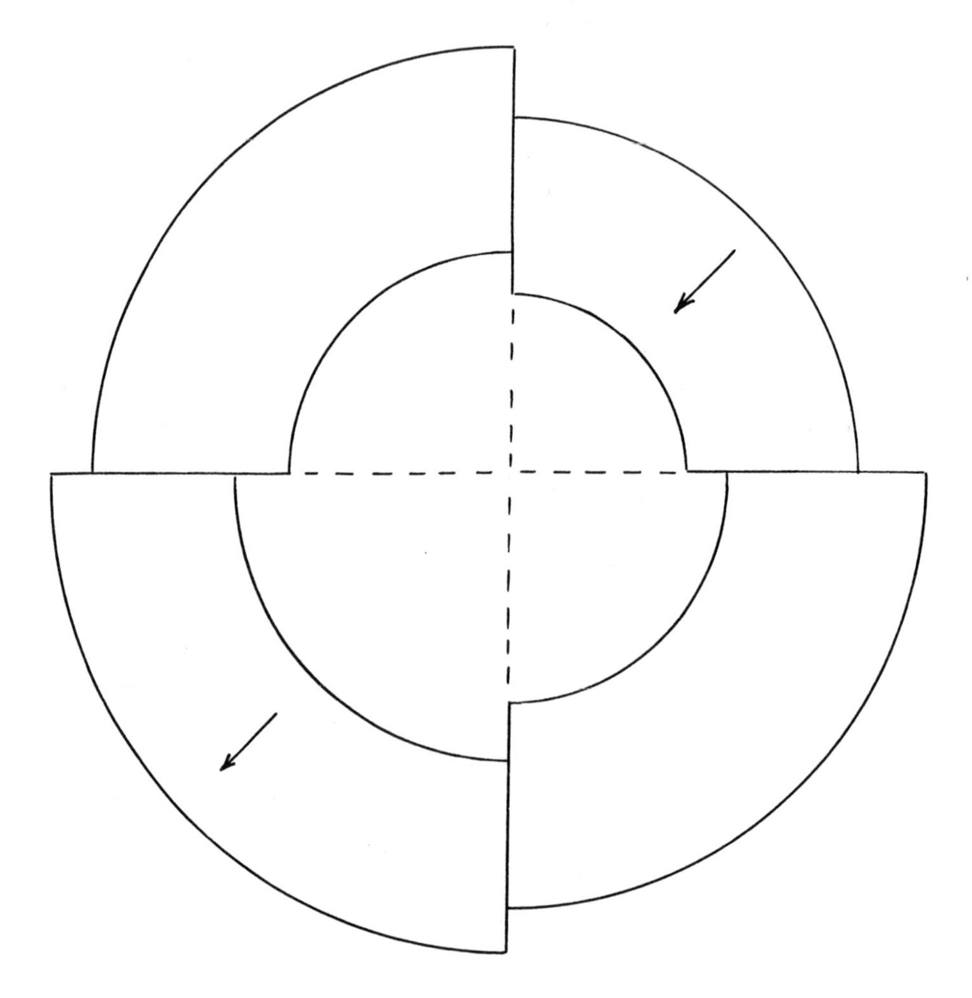

Figure 2-3. The expected basic behavior of the original 5400-km primordial Earth in response to the fundamental driving mechanism of the Tectonospheric Earth Model from the standpoint of radial translatory equilibration. Schematic; not to scale. See text for details.

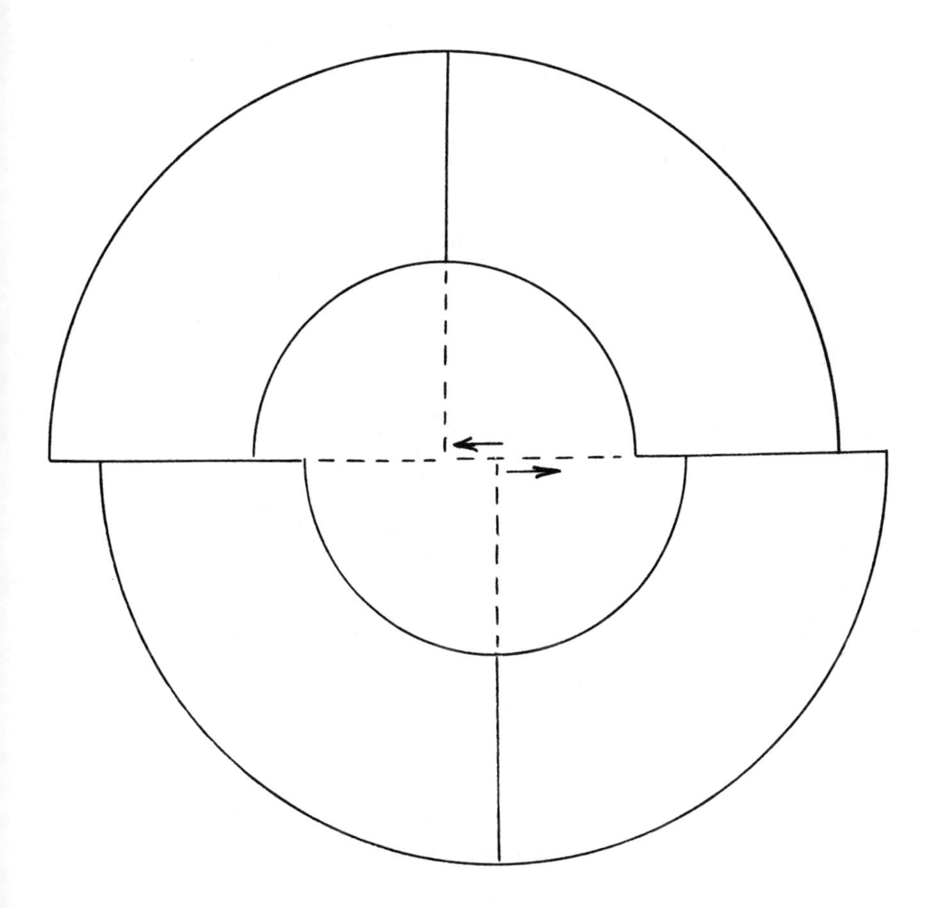

Figure 2-4. The expected basic behavior of the original 5400-km promordial Earth in response to the fundamental driving mechanism of the Tectonospheric Earth Model from the standpoint of transverse translatory equilibration. Schematic; not to scale. See text for details.

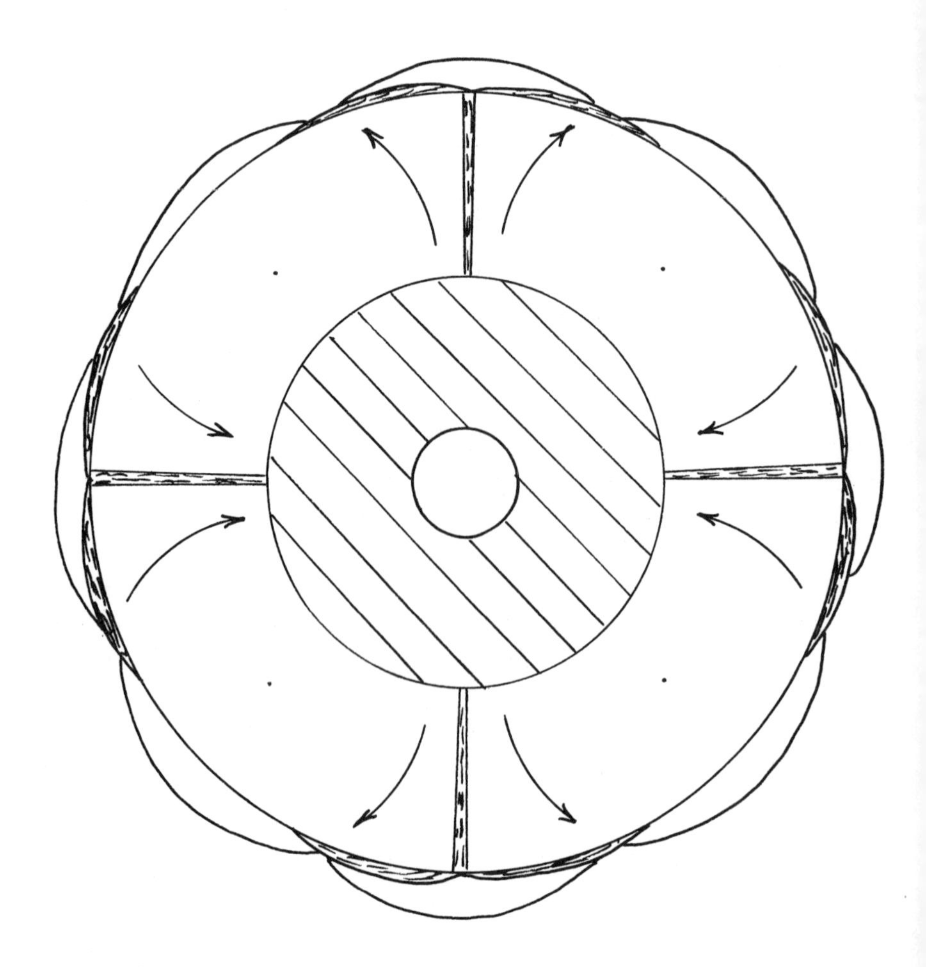

Figure 2-5. The expected basic independent behavior of the individual octants of the original 5400-km primordial Earth in response to the fundamental driving mechanism of the Tectonospheric Earth Model. Schematic; not to scale. See text for details.

has several degrees of freedom with respect to its individual center of gravity. Because of the constraints of the model, this evidences itself mainly as alternate tension and compression along the three mutually-orthogonal fracture planes. This effect is oscillatory, with a long time constant. At any particular time, other things being equal, tension should be found at about half of the octantal contact surfaces with compression at the others. Thus, since each great circle of the basic 5400-km primordial Earth is about 2 pi 5400 = 34000 km long, or 3 x 34,000 = 102,000 km total for all three great circles, then about $\frac{1}{2}$ x 102,000 = 51,000 km of the surficial fracture pattern of the primordial Earth should be in tension while an equal amount is in compression (or neutral). Since the model is continually in the process of equilibrating itself, the locations of the specific segments which are in tension will tend to drift continually along the 102,000-km length of fracture system on the surface of the primordial Earth. Thus, from the standpoint of an external observer, the specific segments in tension would appear to be moving along the 102,000-km linear system in a cyclical but aperiodic manner, since the equilibration of the primordial Earth would, of necessity, be proceeding in a like manner (i. e., cyclically but aperiodically). The rate of such shifting of the tensile segments is of the order of cm/yr.

Segments not in tension at any given time are assumed to be in compression or in a neurtal state. Such segments, whether compressive or neutral, would lie between successive tensile segments and would, therefore, also shift along the 102,000-km subtectonospheric fracture system in a cyclical but aperiodic manner.

Post-Primordial Tectonospheric Accretion.

In evolving the post-primordial Earth from the basic 5400-km primordial Earth, three basic evolutionary modes may be assumed, depending upon the degree to which the 5 terrestrial octants of Earth Prime are fragmentized prior to accretion onto the primordial Earth. These modes are: (1) an octantal-fragment mode; (2) a multiple-fragment mode; and (3) a composite-fragment mode.

The Octantal-Fragment Accretionary Mode.

In the octantal-fragment mode, the 5 octants of Earth Prime are not completely fragmentized prior to accretion onto the basic 5400-km primordial Earth. In this case, parts of each of the 5 octants of Earth Prime might still exist today as unmodified portions of the Earth's tectonosphere, perhaps even as subsurficial portions of presently-existing continental shields. This possibility is considered elsewhere (Tatsch, 1972a: chapter 7).

The post-primordial Earth, in the case of the octantal-fragment mode, evolves essentially from a combination of two processes: (1) the accretion of 5 octants of Earth Prime onto the basic 5400-km primordial Earth; and (2) the differentiation, metamorphism, reworking, igneous activity, and other results of the action of the driving mechanism of the basic model during the past 4.6 billion years (b. y.).

It may be recalled that the driving mechanism of the basic model consists (and has consisted since primordial times) essentially of the resultant of two factors, each of which has been varying in a predictable manner during the past 4.6 b.y.: (1) the potential energy inherent in the geogenetically-disequilibrated shape of the basic 5400-km primordial Earth; and (2) the selectively-channeled energy from the preferential flow of heat, and possibly of volatiles, outward from the three mutually-orthogonal "active" belts formed by the basic octantal fracture pattern of the 5400-km primordial Earth.

Here, as in subsequent discussions and analyses regarding the evolution of the Earth's tectonosphere during the past 4.6 b. y., it is necessary to recall that, in a homogeneous sphere of 6371-km radius, the inertia of the inner 5400-km part normally exeeeds that of the outer 971-km "shell" by at least 60%. Thus, for most purposes, it may be assumed that the disequilibrating inertia of the inner 5400-km part of the model exceeds (and has always exceeded) the equilibrating inertia of the outer 971-km part by a factor of at least 60%. For a non-homogeneous, differentiated body such as the actual Earth, this inertial factor would, of course, normally be even greater than 60%. The significance of this figure is more obvious when some of the mechanics and geometry of the model are discussed in connection with some of the observed behavioral patterns of the actual Earth, such as has been done by the author (Tatsch, 1972a: chapter 8, for example).

The Multiple-Fragment Accretionary Mode.

The multiple-fragment accretionary mode differs from the octantal-fragment mode primarily in the degree to which the 5 "terrestrial" octants of Earth Prime were fragmentized prior to accretion onto the basic 5400-km primordial Earth. For example, if the degree of pre-accretion fragmentation was fairly complete, then it is likely that none of the actual surficial rocks of the present continental shields would now be identifiable with any unmodified portions of the original octants of Earth Prime. This is discussed in greater detail elsewhere (Tatsch, 1972a: chapters 7 and 10).

The Composite-Fragment Accretionary Mode.

The composite-fragment accretionary mode may be considered as an intermediate mode lying between the octantal-fragment mode and the multiple-fragment mode. Depending upon the specific degree of fragmentation assumed for the octants of Earth Prime prior to accretion onto the basic 5400-km primordial Earth, this mode occupies any one of a myriad of intermediate positions within the entire continuum, or envelope, bounded by the octantal-fragment mode and the multiple-fragment mode as extremes.

By extending the limits of the mean, the composite-fragment mode may be made to include one or both of the extremes, thereby collapsing the three modes into two or one, respectively, In this analysis, the three modes will be treated as separate and distinct modes. In the last analysis, any salient differences in the Earth models evolved by the three modes are virtual rather than real; but, by handling the three modes separately, the analysis is more complete and perhaps easier to follow. For this reason, subsequent discussions will assume that there are actually three modes rather than a single mode spread across an entire spectrum, or envelope, including the end points, of the continuum defined by the two extreme modes. The efficacy of this analytical approach is perhaps more obvious when the present continental structures are analyzed in terms of the most likely size of the post-primordial subcratonic blocks that existed within the Earth's tectonopshere, 4.6 b. y. and 3.6 b. y. ago, as well as during the other critical times in the orogenic-cratonic evolution of the Earth's tectonosphere (See, e. g., Tatsch, 1972a: chapters 7 and 15).

The Post-Primordial Earth.

The post-primordial Earth, as derived from the dual primeval planet hypothesis, consists essentially of two parts: (1) an actively-equilibrating, or mobile, 5400-km primordial Earth, differentiated, and fractured basically into octants by three mutually-orthogonal central planes; and (2) an overlying tectonosphere, consisting essentially of the remnants of 5 fragmented octants of Earth Prime accreted onto the 5400-km primordial Earth to a depth of about 1000 km, as modified continually by the driving mechanism of the basic model. Since the primordial Earth loses its identity by becoming incorporated into the post-primordial Earth, it will henceforth be referred to as the Earth's subtectonosphere, or the subtectonospheric portion of the model.

Fig. 2-6 is a schematic sketch of the post-primordial Earth, showing the earliest effects of both igneous and accretionary activity within the tectonosphere. Accretionary activity is indicated schematically at A; igneous activity at I. The octantal blocks of the subtectonosphere are marked S.

Salient Implications of the Basic Tectonospheric Earth Model.

At this point, it is well to list some of the salient implications of the basic Tectonospheric Earth Model in order that the reader may obtain an overall perspective of their nature and scope:

a. Since the 1000-km accretionary material was added to the primordial Earth over a considerable period of time, it follows that various accreted fragments, either individually or in global and regional "shells", would have retained considerable freedom of motion, both with respect to the other accreted fragments and with respect to the basic 5400-km subtectonosphere. Also, it follows that such freedom of motion would most likely be detectable along various planes, surfaces, and shells of preferential rheidity. This is discussed elsewhere (Tatsch, 1972a); there some of these effects are identified as "wave guides", "low-velocity zones", and similar features and phenomena within the tectonosphere. In this connection, it is well to recall that, at a rate of 1 m/yr, a depth of 1000 km is accreted in 1 m. y.; at a rate of 1 cm/yr, it is accreted in 100 million years (m. y.).

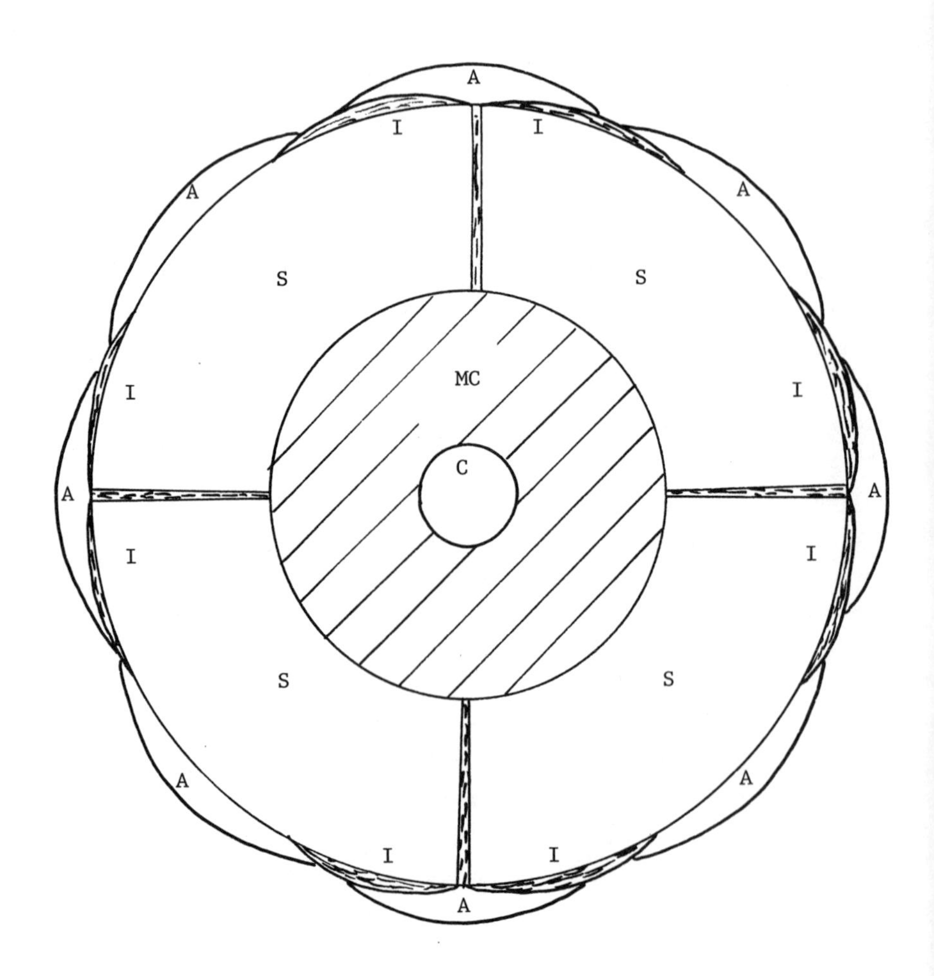

Figure 2-6. The post-primordial Earth, showing the combined effects of primeval igneous and accretionary activities. Accretionary activity is indicated by A; igneous activity by I. The octantal blocks of the subtectonosphere are marked S. MC is the molten core. C is (probably) a small central solid core. Schematic; not to scale. See text for details.

b. The Earth's fracture system consists essentially of the projections, through the tectonosphere, of the basic primordial fracture system of the subtectonosphere. Since the tectonosphere is not homogeneous, such projections are not normally simple, radial projections but somewhat more complex. Other things being equal, the probable error in locating the surficial position of an "energy source" projected from a point on the subtectonosphere through a vertical thickness of 1000 km of average tectonospheric material is roughly 1000 x pi/4, or about 785 km. The significance of this "probable error" figure, as well as the causes for, and magnitudes of, variations in its values are discussed in another work (Tatsch, 1972a; chapter 6), where "cones of activity" are considered in connection with the mechanics and geometry of the evolution of the Earth's tectonosphere during the 4.6 b. y. that the Earth is believed to have been in existence.

c. The Earth's basic deep-fracture system is continuous and worldwide in both the inner (subtectonospheric) and outer (tectonospheric) parts of the model. The latter is basically a consequence of the former, within the constraints of the driving mechanism and the octantally-fractured geometry and mechanics inherent in the model. From this, it appears that the Earth's deep trenches should form a global pattern identifiable with the three mutually-orthogonal great-circular belts of the subtectonosphere. This is discussed elsewhere (Tatsch, 1972a: chapters 6, 7, 8, and 18).

d. "Active" regions (new islands, new mountains, seismic ridges, new volcanoes, etc.) lie roughly above the basic fracture planes of the inner, subtectonospheric part of the model. Due to the effects of planes, surfaces, zones, and shells of preferential rheidity and heat flow, these "active" regions would not, of course, be expected to lie directly above the subtectonospheric fracture system (Tatsch, 1972a: chapter 9).

e. "Inactive" or "fossil" regions (older mountains, older islands, older volcanic belts, seamounts, aseismic ridges, older mineralization belts, etc.) are identifiable with former positions of projections from the subtectonospheric fracture system. To an external observer, these fossil, or inactive, regions appear to have "drifted", in time, with respect to the presently active regions. Normally, such "drifting" involves both translation and rotation, where neither is necessarily linear with respect to time. Also, it might be more accurate to say that the subtectonospheric fracture system, rather than the surficial manifestations thereof, have "drifted", or perhaps even more accurately that both the surficial manifestations and the basic subtectonospheric fracture system have drifted, depending upon whether one wishes to assume a fixed or mobile datum during the analysis. In either case, of course, the results are the same. Also, according to the basic model, the "fossil" features are spatially subparallel to, and temporally derivable from, the "active", or present, features by backward extrapolation procedures. Needless to say, such spatial and temporal relationships are not necessarily linear with respect to time. This is discussed in greater detail elsewhere (Tatsch, 1972a: chapters 7, 8, and 9).

f. Other things being equal, the age of any particular surficial manifestation of the Earth's internal behavior (such as a mountain range, a crustal rift-ridge system, a volcanic belt, etc.) may be approximated roughly by measuring the <u>gross</u> distance by which the particular surficial manifestation has "drifted" from the present position of its genetic subtectonospheric fracture plane. Thus, it is not too surprising that the Appalachian-Caledonian (Upper Paleozoic) orogeny is geographically removed about 30° from the surficial trace of the nearest sub-tectonospheric fracture plane (i. e., Galapagos-Gibraltar), because a drift of 1 cm/yr is roughly equivalent to about 30° per 300 m. y. This is discussed in greater detail elsewhere (Tatsch, 1972a: chapters 7, 8, 9, 10, 12, 13, 16, 17, 18, and 20).

g. The oldest geological features (or surficial manifestations of the Earth's internal behavior), such as Precambrian shields, presently occupy positions identifiable with the basic fracture system of the subtectonosphere. Specifically, they are either: (1) identifiable with the centers of the surficial octants of the subtectonosphere, if stable; or (2) identifiable with the edges of the surficial octants of the subtectonosphere, if they are unstable, active, or in the process of rifting. An example of the latter might be the South American-African "shield" 200 m. y. ago, which has since been rifted and separated as shown schematically in Fig. 2-7. A more-detailed discussion of this is contained elsewhere (Tatsch, 1972a: chapter 17).

h. Likewise, the newest surficial features of the Earth's internal behavior would be expected to be located near the surficial traces of the three mutually-orthogonal subtectonospheric fracture planes projected to the surface of the Earth.

i. One of the several "unbalanced" conditions existing in the post-primordial Earth resulted from the fact that portions of the Earth's surface must, at various times during the history of the post-primordial Earth, have become squarely juxtaposed above the surficial manifestations of the tectonospheric fracture system. Among the consequences of such a situation would be

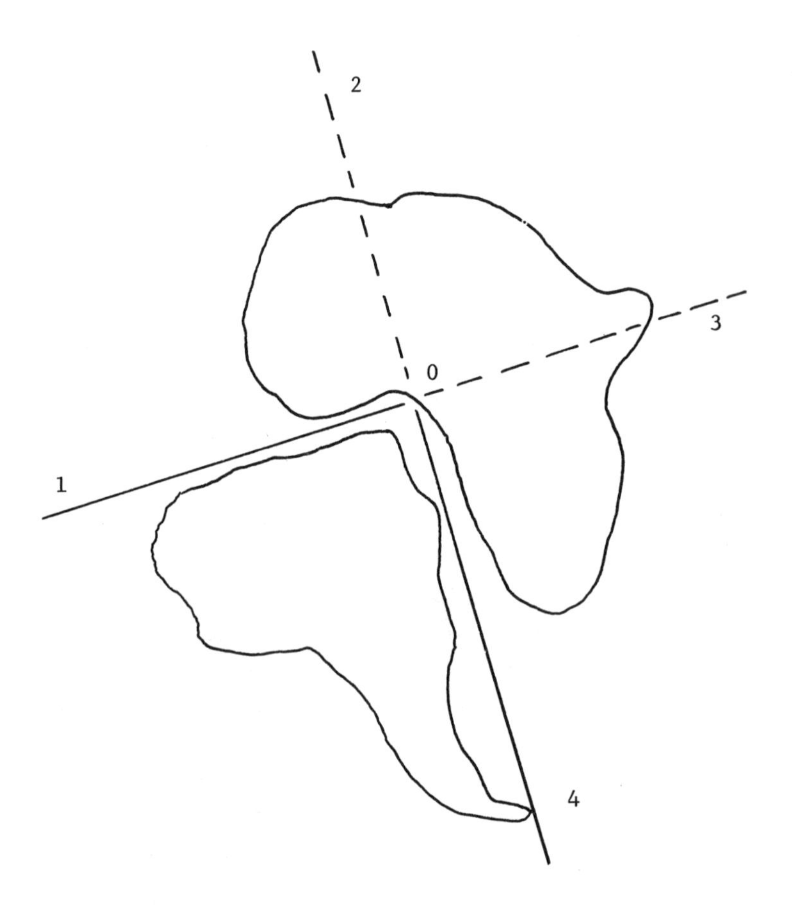

Figure 2-7. Schematic sketch showing two seismotectonomagmatic wedge-belts of activity of the Tectonospheric Earth Model severing Africa from South America during the Mesozoic. The belt segments 1-0 and 4-0 were "tensile" during that part of the Mesozoic; the segments 2-0 and 3-0 were compressive or neutral. See text for details. (Adapted from Tatsch, 1972a).

sea-floor spreading and continental drift (Tatsch, 1972a: chapters 16 and 17).

j. The stress release and other equilibrating effects from the potential energy inherent in the geogenetically-disequilibrated shape of the subtectonosphere are the basic cause of the Earth's deep seismicity (Tatsch, 1972a: chapter 8).

k. Mantle convection may be motivated by the preferential flow of heat along the patterns provided by the subtectonospheric fracture system. Under this concept of the model, mantle convection is a manifestation of the driving mechanism of the basic model. Thus, unlike most other hypotheses, the Tectonospheric Earth Model does not invoke mantle convection as a motive force for sea-floor spreading and continental drift. Rather, mantle convection is derivable from the basic model as an inherent manifestation of the same driving mechanism that drives sea-floor spreading, continental drift, and other omniductive phenomena on the surface of the Earth. The mechanics and geometry of these phenomena, as viewed by the model, are discussed in another work (Tatsch, 1972a: chapters 16 and 18).

l. Igneous activity, both intrusive and extrusive, is motivated by the preferential flow of heat, and possibly of volatiles, outward from the subtectonospheric fracture system (Tatsch, 1972a: chapter 10; 1973a: chapter 4).

m. The preferential flow of heat outward from the subtectonospheric fracture system may be employed to furnish the basic motive power for, and a stable framework for maintaining, a geomagnetic dynamo with a long time constant (Tatsch, 1972a: chapter 14).

n. The roughly antipodal locations of the oceans with respect to the continents is basically a consequence of two factors: (1) the radial translatory motion of the subtectonospheric octantal blocks with respect to the center of the model; and (2) hydrostatic adjustment, or isostasy, on a global scale over a long period of time (Tatsch, 1972a: chapter 15).

o. If viewed on a global scale, certain features such as fault lines, rift-ridges, mountain belts, seamount chains, etc., should normally tend to display "parallel" surficial patterns, intersected by transverse features. As a result, other things being equal, they should tend to form surficial manifestations suggestive of pinnate, grid, and arcuate patterns. These reflect both translation and rotation between the subtectonospheric blocks and the surface of the Earth, most probably along planes, shells, surfaces, and zones of preferential rheidity within the tectonosphere (Tatsch, 1972a: chapters 6 and 18).

The above are a few of the "first order" implications of the basic Tectonospheric Earth Model. The details of these and other implications of the model are discussed elsewhere (Tatsch, 1972a). Higher order implications of the model are the subject of other publications (See, e. g., Tatsch, 1973a through 1973j; 1974b; 1975a; 1975b; 1975c).

<u>Global Surficial Manifestations of the Earth's Internal Behavior Predictable from the Basic Tectonospheric Earth Model</u>.

One of the postulates of the basic Tectonospheric Earth Model is that almost all surficial features and phenomena of the Earth, both past and present, are fundamental consequences of a single, deep-seated driving, or causal, mechanism that has resided within the Earth since the formation of the primordial

Earth, about 4.6 b. y. ago. The details of exactly how such surficial manifestations evolved as a consequence of the behavior of the Earth's tectonosphere during the past 4.6 b. y. is beyond the scope of this book, but many of these details have been discussed elsewhere (See, e. g., Tatsch, 1972a). It may be well to list some of these surficial manifestations of the Earth's internal behavior, in order that the reader may have an appreciation for the general nature and scope of those surficial manifestations that, according to the model, are internally-induced in a predictable manner. These include: global and regional tectonism and orogenesis; sea-floor spreading and continental rifting; global and regional heat-flow patterns; the Earth's magnetic field and its variations, including cyclical but aperiodic reversals in polarity; the Earth's gravity field; volcanic activity; mountain building and related geosynclinal activity; the development of continental shields, mobile belts, fracture systems, rifts, and ridges; continental splitting, separation, and drifting; earthquakes and global and regional seismicity patterns; gravity anomalies and global and regional geoidal patterns; heat-flow anomalies and global and regional thermal patterns; electrical conductivity anomalies; global seismotectonomagmatic and mineralization belts; the Earth's plate-tectonic and omniductive behavior; etc.

At this point it is well to examine one of the surficial manifestations of the Earth's internal behavior, in an attempt to provide a tentative answer to the compound question: Is there any observational evidence that the Earth might, in fact, be representable by the Tectonospheric Earth Model; and, if so, how might the present positions of the eight hypothetical subtectonospheric blocks be identified or located?

To answer this compound question, consideration may be given to a preliminary analysis of any one of several sets of applicable observational data regarding the gross surficial characteristics of the Earth. For example, one such set of data is that showing the locations of the active volcanoes. Since volcanoes appear to be thermally induced and since they are surficial manifestations of tectonospheric activity of some type, they might be expected, according to the model, to lie along segments of global belts defined roughly by the surficial traces of the projections from the three mutually-orthogonal great-circular belts forming the subtectonospheric fracture system.

Therefore, as a first approximation to the location of the surficial manifestations of the subtectonospheric fracture system, we may perform a three-dimensional least-squares analysis of the distribution of the Earth's active volcanoes. Allowing for the probable error expected from the 1000-km projection through the heterogeneous tectonospheric material, we should expect almost all of the Earth's active volcanoes to lie within a surficial distance of about 1000 x $\pi/4 \doteq 785$ km from segments of the radial projections of the subtectonospheric fracture pattern.

When the suggested theee-dimensional least-squares analysis is done, it is found (Tatsch, 1964a) that approximately 93% of the active volcanoes fall within the belts predicted by the model. Specifically, the active volcanoes appear to form segments of three great-circular belts intersecting at the following 6 points, which may be designated tentatively as the "basic tectonospheric points" of the present surficial manifestitions of the basic Tectonospheric Earth Model: 55°N, 165°W (Aleutians); 55°S, 15°E (Bouvet); 5°S, 85°W (Galapagos); 30°N, 5°E (Gibraltar); 5°N, 95°E (Bengal); and 30°S, 175°W (Kermadecs).

Because the probable error of the projection from the subtectonosphere to the surface of the Earth is about $785/111 \doteq 7^{\circ}$, no purpose is served in attempting to locate the coordinates of the tectonospheric points more "accurately" than about 5° in either latitude or longitude. The geographic names (Aleutians, Bouvet, etc.) that are shown above in parentheses are intended for associative identification of the general areas of the points, rather than as exact geographical descriptions of the points. The associative names will normally be used in subsequent discussions as a matter of convenience; but it should be understood that these points are projections of 6 other points that form a pattern on the surface of the subtectonosphere, 1000 km beneath the surface of the Earth.

For the sake of simplicity, the first two points (Aleutians and Bouvet) may be referred to as the "tectonospheric poles" of the present surficial manifestations of the basic Tectonospheric Earth Model; the other four points (Galapagos, Gibraltar, Bengal, and Kermadecs) may be considered as defining roughly the "tectonospheric equator" of the present surficial manifestations of the basic Tectonospheric Earth Model. Fig. 2-8 shows the above tentative "tectonospheric coordinate system" of the present surficial manifestations of the basic Tectonospheric Earth Model superimposed upon a Mollweide projection of the salient geographical features of the actual Earth. In the figure, the three mutually-orthogonal great circles (ACDFA, ABDEA, and CBFEC) are the radial projections of three similar great circles defining the subtectonospheric fracture system. Each of the intersections at A, B, C, D, E, and F is an orthogonal intersection, and each of the 8 triangles (CBA, CBD, CED, CAE, FEA, FDE, FBD, and FBA) is an equilateral spherical triangle (i. e., with all sides and angles equal to 90°). On the surface of the tectonosphere, 90° equals to about 10,000 km. On the surface of the subtectonosphere, 90° equals about 8500 km.

The great circle, ABDEA, also marked QQQQ in Fig. 2-8, is a projection of the subtectonospheric equator and it defines roughly the tectonospheric equator of the present surficial manifestations

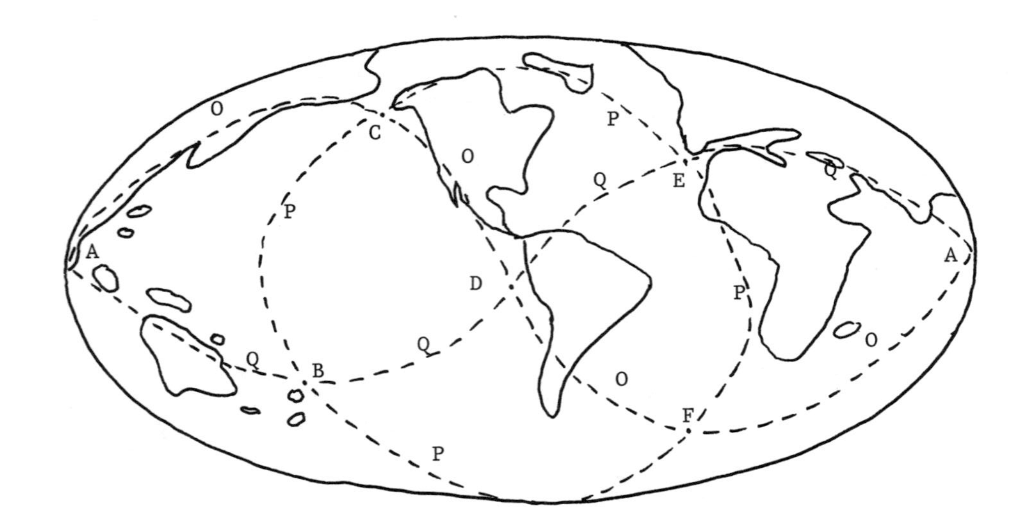

Figure 2-8. Tentative "tectonospheric coordinate system" of the present surficial manifestations of the basic Tectonospheric Earth Model superimposed upon a Mollweide projection of the salient geographical features of the actual Earth. See text for explanation of symbols.

of the basic Tectonospheric Earth Model. The great circle, CBFEC, also marked PPPP, is a projection of the subtectonospheric prime meridian and it defines roughly the tectonospheric prime meridian of the present surficial manifestations of the basic Tectonospheric Earth Model. Similarly, the great circle, ACDFA, also marked OOOO, is a projection of the subtectonospheric orthogonal meridian.

A tectonospheric coordinate system based on the above is particularly useful as a base for performing harmonic analyses of the Earth's surficial manifestations. According to the dual primeval planet hypothesis, the Earth's surficial manifestations are tectonospherically induced and, therefore, "harmonic" with respect to the tectonospheric coordinate system, rather than with respect to the geographic coordinate system normally used for harmonic analyses when using other models.

The Tectonospheric Earth Model as a Supplementary Tool in the Exploration for Minerals.

Briefly stated, the Tectonospheric-Earth-Model approach to mineral exploration is based upon a straightforward premise: if one knew today where all the Earth's minerals were 4.6 b. y. ago, and if he could define the geometrical, mechanical, thermal, and chemical behavior of the Earth during the past 4.6 b. y., then he should be able to predict the present locations of the Earth's mineral deposits (Tatsch, 1973a).

The following chapters discuss how the Tectonospheric Earth Model may be used to define the complex geometrical, mechanical, thermal, and chemical interrelations that exist among igneous activity, folding, faulting, uplifting, subsidence, and other omniductive geophenomena that have contributed to the origin, evolution, and present characteristics of uranium deposits.

Chapter 3

THE GEOMETRICAL, MECHANICAL, THERMAL, AND CHEMICAL BEHAVIOR OF THE EARTH DURING THE PAST 4.6 BILLION YEARS

The Tectonospheric Earth Model differs from simple "plate-tectonics" concepts in two fundamental respects: (1) it extends the applicable spatio-temporal domain to a depth of at least 1000 km and to a span of about 4.6 b. y.; and (2) it provides a single, long-lived, deep-seated, global driving mechanism that has been motivating the Earth's internal behavior during the 4.6 b. y. that the Earth is believed to have been in existence (Chapter 2; Tatsch, 1972a). It is well to consider, at this point, some of the aspects of these two extensions that have been provided by this model.

For ease of presentation, three aspects of the Earth's surficial features and phenomena may be considered: (1) a global view of the Earth's present surficial features; (2) the most probable surficial features of the proto-Earth 4.6 b. y. ago; and (3) the evolution of the Earth's present surficial features from those of the proto-Earth during the past 4.6 b. y. Because the Earth's evolutionary period appears to have been interrupted about 3.6 b. y. ago, the total time period will be considered in 2 parts: (1) 4.6 b. y. ago to 3.6 b. y. ago; and (2) 3.6 b. y. ago to the present.

A Global View of the Earth's Present Surficial Features and Phenomena.

One of the theses of the Tectonospheric Earth Model is that the Earth's surficial features and phenomena are a reflection of its internal behavior. It is well to review some of these features and phenomena as they are interpreted in terms of the Tectonospheric Earth Model concept:

a. Orogenic-cratonic structure of the continents. The

orogenic-cratonic structure of all continents appears to be more-or-less the same (Tatsch, 1972a: chapter 7). What is seen, in general, is an old Precambrian shield successively surrounded by long rectilinear orogenic, geosynclinal, or "seismotectonomagmatic" belts, often broken or otherwise modified by overprinting by subsequent seismotectonomagmatic activity, with at least 3 orogenies fairly well preserved in at least some continents; e. g., the Appalachian, Innuitian, and Cordilleran orogens of the Canadian shield. Unanswered are questions as to why the orogenic belts are rectilinear and of different ages within any one continent but appear to to extend worldwide, with their activity having occurred in cyclical-but-aperiodic episodes for at least the past 3.6 b. y., and with concentrations of this activity having occurred at about 0.9 b. y. intervals, or at sub-multiples thereof. Also not understood is why the spatio-temporal relationships within all orogenic belts are remarkably similar, regardless of where found or to which period they belong. However, in spite of these unanswered questions, the orogenic-cratonic structure of the continents suggests a remarkably orderly behavior over an extremely long period of time, perhaps as long as 3.6 b. y. or even 4.6 b. y. This, in turn, suggests that the continental structures are surficial manifestations of a long-lived, deep-seated, global driving mechanism.

b. <u>The Earth's seismicity and earthquakes</u>. When considered as a surficial manifestation of the Earth's internal behavior, earthquakes suggest that the Earth's internal behavior is controlled by a global mechanism completely independent of continental margins, island arcs, and similar features, except insofar as they may also be surficial features of the same mechanism (Tatsch, 1972a: chapter 8).

c. <u>Heat-flow and other geothermal activity</u>. When viewed on a global scale, it appears that the Earth's internal behavior reveals itself through thermal manifestations on the surface. The exact correlation between internal behavior and surficial manifestations of the thermal Earth is not clearly understood, perhaps because of the paucity of global heat-flow data. If a generalization may be made, it is that regions of modern orogenesis correlate with high heat-flow rates (Tatsch, 1972a: chapter 9).

d. Intrusive and extrusive activity. When viewed on a global scale, intrusive and extrusive activity, of both the mid-ocean-ridge type and the Alpine-ultramafic-belt type, may be considered to be surficial manifestations of a single, global mechanism characterized by cyclical-but-aperiodic activity along long, linear belts of deep-seated control. Closer scrutiny reveals that intrusive and extrusive activity forms a complete spatio-temoral spectrum, or continuum, within each cycle of orogenic activity whether of the mod-ocean-ridge type or of the Alpine-ultramafic-belt type (Tatsch, 1972a: chapter 10).

e. Mountains and geosynclines. The relationship that exists between mountains and geosynclines is not completely understood. It appears, however, that, when geosynclines and mountains are analyzed simultaneously on a global scale, then their origin and evolution appear to be controlled by a single, global mechanism deep within the Earth (Tatsch, 1972a: chapter 12).

f. The Earth's gravity field and the shape of the geoid. On the basis of present evidence, it appears that the Earth's gravity field and the shape of the geoid are surficial manifestations of the internal behavior of the Earth during the past 4.6 b. y. However, it appears also that mass redistributions lag behind the forces causing the Earth's internal behavior. The shape of the geoid should correlate, therefore, with the tectonospheric driving mechanism rather than with purely surficial manifestations of the Earth's internal behavior (Tatsch, 1972a: chapter 13).

g. Geomagnetism and polarity reversals. When the Earth's magnetic field is considered over a long period ot time, several tentative conclusions may be drawn: (1) the geomagnetic dynamo has evolved, and is presently driven, by the same global causal mechanism that has produced the Earth's internal behavior during the past 4.6 b. y.; (2) the geomagnetic field correlates, now and in the past, with global characteristics of the Earth's driving mechanism; and (3) the cyclical-but-aperiodic polarity reversals are definable in terms of the geometry and mechanics of the Earth's driving mechanism (Tatsch, 1972a: chapter 14; 1974c). On this basis, it appears that, when viewed on a global scale, then geomagnetism and polarity reversals are surficial manifestations of the Earth's internal behavior.

h. Continents and oceans. Viewed on a global scale and over a long period of time, it appears that oceans and continents are surficial manifestations of the Earth's internal behavior (Tatsch, (Tatsch, 1972a: chapter 15).

i. Crustal rifting and sea-floor spreading. Crustal rifting (and sea-floor spreading as a subsidiary phenomenon thereof) appears to be a surficial manifestation of the Earth's internal behavior. Furthermore, such rifting (and spreading) appears to occur when and wherever the Earth's internal behavior causes crustal tension (Tatsch, 1972a: chapter 16).

j. Plate tectonics and related matters. If plate motions are surficial manifestations of the Earth's internal behavior, then the origin, evolution, and present characteristics of the plates should be correlatable with certain features and phenomena of the Earth's internal behavior (Tatsch, 1972a: chapter 18).

k. Continental drift and polar wandering. When viewed on a global scale over a long period of time, it appears that continental drift and polar wandering are surficial manifestations of the Earth's internal behavior, during possibly the entire existence of the Earth, but that the geometry and mechanics of the Earth's interior and the nature of the driving mechanism are difficult to define in terms of simple plate-tectonics concepts (Tatsch, 1972a: chapter 17).

In analyzing the present surficial features of the Earth that might have been caused by possible internal-behavior patterns of the Earth during the past 4.6 b. y., it is well to consider briefly the possible surficial configuration of the Earth 4.6 b. y. ago and at various critical times since then.

The Earth's Most Probable Surficial Features 4.6 b. y. Ago.

The configuration of the primordial Earth's most probable surficial features depends upon exactly which hypotheses are adopted to analyze the evolution of the Earth from the solar nebula (Tatsch, 1972a: chapter 1) and to explain the evolution of the Earth-Moon system (Tatsch, 1974a). If it is assumed that the Earth accreted from the solar nebula and that the Moon formed in the vicinity of (but not part of) the Earth, then the Earth's surface would probably have been devoid of any significant topographic features other than those that would be genetic to an accretionary process. Thus, the Earth's surface would have presented a more-or-less homogeneous aspect 4.6 b. y. ago, both from the standpoint of topography (i. e., smooth) and from the standpoint of composition (i. e., more-or-less the same) over the entire surface.

If, on the other hand, it is assumed that the Earth's surficial composition and/or topography were not homogeneous 4.6 b. y. ago, then such features would most probably not have survived to this date without some internal of external rejuvenation at least on one occasion since then. For the purpose of this analysis, it may be assumed that the Earth's surface 4.6 b. y. ago was devoid of any features other than those that would have been genetic to an accretionary formation from the solar nebula at that time (Tatsch, 1974a).

The Earth's Most Probable Surficial Features 3.6 b. y. Ago.

If it is assumed that the Earth had no appreciable surficial features 4.6 b. y. ago, when were the surficial features first acquired, and what caused them to form? Were they acquired through external or internal behavior? If external, what was it? If internal, why did it occur when it did, rather than sometime eralier or later?

If the Earth's surficial features resulted from external sources or causes, during the period 4.6 b. y. to 3.6 b. y. ago, then such features would not now be active but would resemble the Archean shields and would have an age of about 3.6 b. y.

If the Earth acquired its surficial features from internal sources or causes, during the period 4.6 b. y. to 3.6 b. y. ago, then such features would not now be active unless the same causative internal hehavior had been operating ever since 3.6 b. y. ago. In that case, it would be reasonable to assume that the internal mechanism has been operating more-or-leas continuously during the past 4.6 b. y. (because nothing internal to the Earth would have been available to activate it 3.6 b. y. ago).

In any case, it appears that the Precambrian shields existed 3.6 b. y., ago, but not necessarily in their present forms and locations. Also, it appears that these shields have existed in one or more locations and forms ever since then, without having been severely heated or shocked except at their peripheries.

The Evolution of the Earth's Present Surficial Features.

Regardless of whether it is assumed that the Earth's surficial features of 3.6 b. y. ago were externally or internally induced, we must consider certain questions about what has happened to the Earth's surface since then. For example, have there been significant externally-induced catastrophes that migth have altered the Earth's surficial features during the past 3.6 b. y.? If so, what were they, when did they occur, and what caused them? If no such external influences have operated during the past 3.6 b. y., is it reasonable to assume that all modifications to the Earth's surface since then were induced by the Earth's internal behavior?

To answer these and related questions about the evolution of the Earth's present surficial features, global-scale analyses must be made regarding both the present and past internal behavior of the Earth. Several international scientific bodies have been devoted to such analyses, e. g., the International Upper Mantle Commission and the Interunion Commission on Geodynamics (Drake, 1970, 1973). Other studies have been made on a national or institutional basis (See, e. g., Anderson, 1975; Bullen, 1974a, 1974b; Irving and McGlynn, 1975; Kaula, 1975; Berg and Sutton, 1975; Oyarzun, 1975; Cabre, 1975; Schilling and Bonatti, 1975; Yoshii, 1975; Vogt, 1975; Richter and Parsons, 1975; Froidevaux and Schubert, 1975; Rosendahl et al., 1975; Oldenburg and Brune, 1975. See also the many earlier references that are cited in Tatsch, 1975a: pages 78 and 79).

When the Earth is considered as a long-lived, global entity, the results of the above, and other, analyses indicate that most past and present geophenomena are manifestations of the Earth's internal behavior during the past 4.6 b. y., but that the geometry and mechanics of the causative internal behavior and the nature of the ultimate driving mechanism remains to be defined. It is the writer's hypothesis that the Tectonospheric Earth Model, derived from a dual primeval planet hypothesis (Tatsch, 1972a) provides some answers for the undefined geometry and mechanics and for the nature of the Earth's internal driving mechanism.

Chapter 4

PLATE TECTONICS, OMNIDUCTIVE PROCESSES, AND SEISMOTECTONOMAGMATIC BELTS

It has been known for some time that mineral deposits are often found near plate boundaries and near other surficial vestiges of the Earth's internal behavior (Tatsch, 1972a). One of the purposes of this chapter is to attempt to generalize this observation into a long-lived deep-seated concept that may best be expressed in terms of "omniductive processes" and "seismotectonomagmatic belts". Such a concept forms one of the bases for using the Tectonospheric Earth Model as a supplementary tool in the exploration for uranium deposits.

Global Seismotectonomagmatic Belts.

A close scrutiny of the Earth's surface reveals that certain elongate belts, characterized by varying degrees of seismic, tectonic, and magmatic activity, have existed in various regions of the Earth during most of the 4.6 b. y. that the Earth is believed to have been in existence. These belts may be referred to as seismotectonomagmatic belts.

Much evidence suggests that all these seismotectonomagmatic belts are similar, regardless of when and where formed (See, e. g., Marlow et al., 1973, who have considered island arcs, orogenic belts, and certain other tectonomagmatic episodes that fall into the category of this definition). Other recent studies of the various facets of the geometrical, mechanical, thermal, and chemical aspects of some of the seismotectonomagmatic belts, of various ages and in various parts of the world, have been completed. These have been summarized elsewhere (Tatsch, 1975a: pages 90 through 115). They need not be repeated here. Suffice it merely to list some of the Earth's features and phenomena that have been interpreted in terms of the seismotectonomagmatic-belt concept of the Tectonospheric Earth Model, where the following listed page numbers refer to Tatsch, 1975a:

a. The Earth's earliest Precambrian activity: page 90.

b. The evolution of Australia: page 94.

c. The development of the Earth's earliest orogenic belts: page 95.

d. The development of the Earth's sialic crust: page 97.

e. The evolution of mountain belts: page 100.

f. The development of faunal assemblages: page 102.

g. Continent-continent collisions: page 102.

h. Descending lithospheric plates: page 104.

i. Transform faults: page 104.

j. Inter-arc basins: page 106.

k. Omniductive processes and ultramafic rocks in paired belts: page 108.

l. Gravity tectonics: page 108.

m. The observed seismicity of the Baffin Bay region: page 110.

The Ages of Sedimentary Rocks as a Measure of Episodes of Seismotectonomagmatic-Belt Activity.

One measure of the activity of seismotectonomagmatic-belt episodes is provided by the youth of the Earth's sedimentary rocks. Although 2/3 of the Earth's present rocks are sedimentary (See, e. g., Blatt and Jones, 1975), over 1/2 of these sediments are younger than Jurassic (roughly 130 m. y.). It is not surprising that less than 5% of all Precambrian rocks are sedimentary. These observations may be expressed mathematically by saying that the spatio-temproal distribution of sedimentary rocks is lognormal with a half-like of 130 m. y. This distribution approximates that predicted by the Tectonospheric Earth Model (Tatsch, 1972a).

Global Synchroneity of Seismotectonomagmatic-Belt Activity.

The surface of the Earth desplays almost an infinity of examples of the global penesynchroneity of seismotectonomagmatic-belt activity during the past 4.6 b. y. The oldest of these examples are embodied in the networks of vestiges of seismotectonomagmatic-belt activity that crisscross the Archean shields. Among others may be mentioned the global mineral belts of all ages, the present Pacific "ring of fire", the Mediterranean-Indonesian arc, etc. (Tatsch, 1972a).

Mineral Deposits as Interpreted by the Seismotectonomagmatic-Belt Concept.

As indicated earlier, many mineralizations appear to be aligned in belts that are closely associated with plate boundaries and with certain other tectonomagmatic, seismotectonic, and omniductive features of the Earth's surface. The exact nature and cause of this association between mineralization belts and the geologic structure and behavior of the Earth are not known. One of the purposes of this book is to make a preliminary analysis of this association insofar as it pertains specifically to uranium deposits.

A global analysis of the Earth's surface suggests that local, regional, and global metallogenic zoning is produced by (1) different structural environments; (2) crust and mantle heterogeneities; and (3) different histories of evolution (See, e. g., Radkevich, 1972). Closer scrutiny shows that these factors are vestiges of seismotectonomagmatic-belt activity that has occurred within the Earth during the past 4.6 b. y. (Tatsch, 1972a).

Chapters 6 through 10 take a closer look at some of the Earth's uranium deposits as vestiges of the seismotectonomagmatic-belt activity during the Cenozoic, Mesozoic, Paleozoic, Proterozoic, and Archean. Before doing that, however, it is well to take a brief look at the concept of using observations regarding intruso-extrusive activity as a long-lived deep-seated basis for the origin, and evolution of mineral deposits. This is done in the next chapter.

Chapter 5

INTRUSIVE AND EXTRUSIVE ACTIVITY: A LONG-LIVED, DEEP-SEATED BASIS FOR THE ORIGIN AND EVOLUTION OF MINERAL DEPOSITS

Many intrusive and extrusive episodes appear to be comagmatic even when not contemporaneous. In some of these episodes, ultramafic rocks have been wholly or partly intruded after the main metamorphic and deformational phases of the tectonomagmatic episode. In other episodes, this sequence has been reversed. In still others, intrusive and extrusive phases have been pene-contemporaneous with the metamorphic and deformational phases. The Kenoran orogeny (2.73 to 2.33 b. y. ago) displays several distinct igneous, metamorphic, deformational, and faulting phases within the Canadian Shield (See, e. g., Weber and Stephenson, 1973). In other times and places, similar phases occur but not necessarily in this same sequence.

Intrusive and extrusive activity is closely associated spatially and temporally with tectonic activity. The intrusion of voluminous basaltic dikes in the volcanic and batholithic structures often continues before, during, and after deformation, suggesting thereby a continuous process of basic and acidic magmatism contemporaneous with tectonic activity (See, e. g., Strong, 1973). All alkali olivine basalt series are similar to other alkali olivines regardless of whether formed in deep-oceanic, island-arc, or continental environments (See, e. g., Schwarzer and Rogers, 1973).

These and other recent observations suggest that the various types of intruso-extrusive activity, as well as the associated tectonic activity, are motivated by a long-lived, deep-seated, global mechanism that has been operating within the Earth during most of the 4.6 b. y. that the Earth is believed to have been in existence. It is well to examine these and other aspects of the intrusive and extrusive episodes, in an attempt to define these episodes in terms of their constituting a long-lived, deep-seated basis for the origin and evolution of mineral deposits.

Intrusive and extrusive activity has played an important role in the evolution of the Earth's tectonosphere during the past 4.6 b. y. This intrusive and extrusive activity includes volcanic activity, the creation of basaltic traps, both in continental and in oceanic areas, mantle upwelling at crustal rifts, and the emplacement of batholiths, laccoliths, and other intrusive bodies during the past 4.6 b. y.

This book considers intrusive and extrusive eruptives as a single entity. This is done in order to emphasize the concept that a single global deep-seated, long-lived mechanism is responsible for the behavior of both of these types of eruptives. The term "eruptive" is used in its general sense (Challinor, 1964), i. e., including both intrusive and extrusive activities involved in breaking through from one level to another within the tectonosphere. Thus, an eruptive that "breaks through" to a subsurface level is an intrusive; one that breaks through to the surface, an extrusive. An intrusive during one orogeny may become an extrusive through the action of a subsequent orogeny.

Arthur Holmes pointed out many years ago that the three major problems of igneous petrology are: (1) the nature and source of primary magmas; (2) the mechanisms of magmatic intrusion and emplacement; and (3) the processes that have brought about the manifold diversity of igneous rocks. One additional problem may be added to these: (4) the nature and source of the energy for the ascensive mechanism. These four problems have been discussed elsewhere (Tatsch, 1973a: chapter 4).

Intrusive and Extrusive Activity as Interpreted by the Tectonospheric Earth Model.

Most observers agree that present mineral deposits have been derived from materials that have existed within the Earth for at least 3.6 b. y. But there is no general agreement on the evolutionary processes that have led to the localization of minerals. Many layered deposits of metallic minerals were contemporaneous with the formation of the enclosing rock. Other minerals were introduced later, perhaps through some geometrical, mechanical, thermal, or chemical aspect of those particular rocks that favored a preferential path for the mineralizing materials. Still others were redistributed from earlier mineralizations during subsequent tectonothermomagmatic activity. These redistributions, in some cases, caused enrichment; in others, impoverishment (Bain, 1968).

According to the Tectonospheric Earth Model, the origin, evolution, and present characteristics of mineral deposits are closely associated with the Earth's intrusive and extrusive activity, which, in turn, is associated with orogenic activity. These associations may be seen in the evolution of most continents.

The geological history of Africa, for example, bears a close relationship to the origin, evolution, and present characteristics of mineral deposits within that continent. The history of most continents can be studied best as an orogenic-cratonic evolution. Thus, we find that Africa has undergone at least seven major orogenic events, each of which had a profound effect on the origin, evolution, and present characteristics of the mineral deposits within the African continent: (1) Transvaalian, 3 b. y. ago; (2) Shamvaian, 2.8 to 2.5 b. y. ago; (3) Eburnian-Huabian, 1.85 b. y. ago; (4) Kilbaran, 1.1 b. y. ago; (5) Damaran-Katangan (Pan-African), 0.55 b. y. ago; (6) Acadian-Hercynian, 0.40 to 0.20 b. y. ago; and (7) Alpine, 0.18 to 0.02 b. y. ago.

In Africa, as in other continents, the earliest orogens appear to have been less affected by more recent orogenies. Therefore, the mineral deposits within the oldest orogenic belts can be expected to differ from those within more recent orogenic belts. Thus, the older orogenic belts (long since welded into the present cratons) contain Africa's principal deposits of asbestos, chromium, diamonds, gold, and iron. The younger orogens contain Africa's principal deposits of beryllium, cobalt, copper, lead, niobium, tantalum, tin, and tungsten.

According to the Tectonospheric Earth Model, all minerals that have been deposited during the past 4.6 b. y. have originated within an active seismotectonomagmatic belt (Tatsch, 1972a). Thus, Africa's primary deposits of asbestos, chromium, diamonds, gold, and iron originated within seismotectonomagmatic belts that controlled the three earliest major African orogenies, i. e., the Transvaalian, Shamvaian, and Eburnian-Huabian. Similarly, Africa's primary deposits of beryllium, cobalt, copper, lead, niobium, tantalum, tin, and tungsten became concentrated within the seismotectonomagmatic belts that controlled the four more-recent major African orogenies, i. e., the Kilbaran, Pan-African (Damaran-Katangan), Acadian-Hercynian, and Alpine.

Almost all of the Earth's oldest basic seismotectonomagmatic belts have long since become welded to the cratons (and in most cases are now considered to be part of the respective cratons). However, the Earth's present metallogenic provinces attest to the erstwhile activity of these fossil seismotectonomagmatic belts. In some cases, the process by which these erstwhile active seismotectonomagmatic belts became welded to the respective cratons has caused the original deposits within the older fossil seismotectonomagmatic belts to become less clear-cut than is characteristic within the more-recently active seismotectonomagmatic belts. Thus, minerals characteristic of the older seismotectonomagmatic belts occur where large segments of these older seismotectonomagmatic belts have been involved in younger orogenic events. Consequently, a number of asbestos, gold, and iron deposits found near the margins of younger orogens probably owe their present distribution to this later orogenic activity.

In some cases, minerals originally emplaced in fossil seismotectonomagmatic belts have been transported to deposits within younger seismotectonomagmatic belts. One example concerns the diamondiferous gravels that have been transported by the Orange River into Tertiary and Recent sediments. Similar examples may be found in other continents involving different minerals at different times.

Intrusive and extrusive activity, according to the Tectonospheric Earth Model, is associated with the "cones of activity" as well as with the seismotectonomagmatic belts of the model (Tatsch, 1972a: chapter 6). That is, intrusive and extrusive activity due to a given energy source (existing at the surface of the subtectonosphere) would be expected to occur somewhere within a cone of activity centered on the "energy source" as a vertex and intersecting the surface of the Earth in a circle with a radius of approximately 785 km (= 1000 pi/4) in the idealized case (Tatsch, 1972a: figure 6-2e-2, chapter 6).

When local cones of activity are formed for every point on the subtectonospheric fracture system, then the surficial trace of their vertexes consists of 3 mutually-orthogonal great circles, aggregating about 102,000 km in length. The resulting cones of activity form an infinity of cones whose "envelope" may be described as 3 global tectonospheric seismotectonomagmatic belts. At the surface of the Earth, where the radius is 1.18 times that at the base of the tectonosphere, the centerlines of these cones form 3 mutually-orthogonal great circles aggregating 102,000 x 1.18 = 120,000 km (Tatsch, 1972a: chapter 6).

All intrusive and extrusive activity, according to the model, is motivated by the Earth's driving mechanism. Intrusive and extrusive activity resulting directly from the Earth's driving mechanism would normally be expected to occur within the present position of an active seismotectonomagmatic belt; that resulting indirectly from the Earth's driving mechanism may occur anywhere within the Earth without regard to present positions of the active seismotectonomagmatic belts (Tatsch, 1972a).

Because, at any given time, active or effective energy sources may be expected along about 50% of the subtectonospheric fracture system (i. e., along about 51,000 km), intrusive and extrusive activity would be expected to occur within about half of the global seismotectonomagmatic belt system (i. e., along "belts" aggregating about 60,000 km on the surface of the Earth) at any one time. Intrusive and extrusive activity would not necessarily occur simultaneously in all active segments of the global seismotectonomagmatic belts. Also, some types of magmatic activity normally would not accompany certain types of tectonic activity. For example, ultrabasic effusives would not normally accompany compressive tectonics according to the model. Nor would more than one specific type of magmatism be expected to occur at a given time at a given location (Tatsch, 1972a).

Crustal and Upper-Mantle Evolutionary Modes Postulated by the Tectonospheric Earth Model.

Depending upon what basic assumptions are made, the Earth's crust and upper mantle developed, according to the Tectonospheric Earth Model, through any one of three basic modes of accretionary evolution: (1) octantal-fragment, (2) multiple-fragment, and (3) composite-fragment (Tatsch, 1972a: chapter 3). The type and behavior of magmatic activity during the past 4.6 b. y. would have been different in each of these three accretionary modes of tectonospheric evolution. These may be summarized.

In the octantal-fragment mode for the evolution of the Earth's tectonosphere, the 5 "terrestrial" octants of Earth Prime are not completely fragmentized prior to accretion onto the basic 5400-km primordial Earth (Tatsch, 1972a: chapter 3). In such a case, parts of each of the 5 terrestrial octants of Earth Prime might still exist today as unmodified portions of the Earth's present tectonosphere, perhaps even as subsurficial portions of presently-existing continental shields.

If the Earth's tectonosphere evolved through the octantal-fragment mode, the intrusive and extrusive activity during the past 4.6 b. y. would have been somewhat different from that in the multiple-fragment and composite-fragment modes for the evolution of the Earth's tectonosphere. The details of these differences have been discussed (Tatsch, 1972a), but they need not be considered further here.

The multiple-fragment mode differs from the octantal-fragment mode primarily in the degree to which the 5 "terrestrial" octants of Earth Prime were fragmentized prior to accretion onto the basic 5400-km primordial Earth. If the degree of pre-accretion fragmentation was fairly complete, then it is not very likely that any of the actual surficial rocks of present continental shields would now be identifiable with unmodified portions of the original octants of Earth Prime, because intrusive and extrusive activity during the past 4.6 b. y. would have modified these multiply-fragmentized primordial rocks.

The composite-fragment mode may be considered as an intermediate mode lying between the octantal-fragment mode and the multiple-fragment mode. Depending upon the specific degree of fragmentation assumed to have occurred within the octants of Earth Prime prior to accretion onto the basic 5400-km primordial Earth, this mode occupies one of a myriad of intermediate positions within the entire spectrum, or envelope, bounded by the octantal-fragment and multiple-fragment modes as extremes. A similar statement may be made about the modification of these fragments by intrusive and extrusive activity during the past 4.6 b. y. That is, the amount and degree of modification by intrusive and extrusive activity in the composite-fragment mode would be intermediate between that in the octantal-fragment and that in the multiple-fragment mode of tectonospheric evolution.

<u>Intrusive and Extrusive Activity as a Function of Paths of Preferential Heat Flow within the Tectonosphere</u>.

Within the seismotectonomagmatic belts, the nature and location of intrusive and extrusive activity depends largely upon: (1) the nature of the local stress field and (2) the paths of preferential heat flow (Tatsch, 1972a: chapter 6).

If an energy source at some point along the subtectonospheric fracture system consisted of heat, and if the tectonosphere above that point were homogeneous, then it might be a relatively simple matter to determine the most probable paths that the heat might follow through the tectonosphere to the surface of the Earth. Because the tectonosphere (and specifically the seismotectonomagmatic belt) is not homogeneous, the preferential heat-flow path (from the surface of the subtectonosphere to the surface of the Earth) might be quite circuitous in order to circumvent heterogeneities of low thermal conductivity. The exact path followed by the heat would influence the nature and location of the resulting magmatic activity, whether intrusive or extrusive. This, according to the model, accounts for the circuitous routes taken by some feeders, or conduits, to volcanoes and to intrusive bodies (Tatsch, 1972a: chapter 9).

Paths of preferential heat flow within the Earth's tectonosphere are of interest to the origin, evolution, and present characteristics of mineral deposits because most paths of preferential heat flow are closely associated with paths of preferential flow of volatiles and magmas. The importance of these associations may be emphasized by considering them in connection with the origin of mineral deposits.

Thus, we see that magmas appear to be the ultimate source of essentially all the constituents of mineral deposits. Some of the volatile constituents (e. g., water, oxygen, and carbon dioxide) appear to have followed circuitous paths from the magmas, to the hydrosphere, and then to mineral deposits, but their ultimate source appears to be the magmas.

Magma crystallization may be analyzed in three stages: (1) initial, (2) intermediate, and (3) final. Mineral deposits fall, likewise, into one of these categories, depending upon what stage of crystallization is currently applicable. Thus, many metallic oxides, sulfides, and native metals are representative of the initial stage of magma crystallization. Other minerals, generally silicic, are representative of the intermediate stage of magma crystallization. Still other minerals, highly silicic, are representative of the final stage of magma crystallization.

During the initial stage of magma crystallization, the abstraction of the early crystallization products leaves a normally silicic fluid enriched in volatiles. These tend to collect in the upper part of the magma chamber until a path of preferential escape develops, permitting them, under suitable conditions, to form contact-matasomatic mineral deposits.

Toward the close of the magma solidification, some of the highly silicic accumulated fluids may pass through paths of preferential escape to form pegmatites. These fluids are accompanied or followed by aqueous solutions containing the compounds from which the ores are deposited by replacing the pegmatitic minerals. Upon final consolidation of the magma, the residual fluids are ejected along paths of preferential flow (i. e., of negative pressure gradient). From these, nearly all epigenetic mineral deposits are formed.

As the fluid cools, the gaseous constituents form magmatic liquids. These, upon nearing the surface, form hydrothermal solutions, either with or without the addition of meteoric waters. Some alteration occurs as these solutions traverse paths of preferential flow through the various wall rocks. Many become the alkaline solutions from which ores are deposited.

The paths of preferential flow may be cracks, joints, bedding planes, rock pores, and other openings that might have been produced by the geometrical, mechanical, thermal, and chemical behavior of the Earth's tectonosphere during the past 4.6 b. y. The great variety of wall rocks involved in the various conditions of temperature and pressure determines the exact locations for the hypothermal, mesothermal, and epithermal deposits.

Subsequent processes may alter the initial deposits. For example, weathering may release minerals for transport to sedimentary basins. Organic processes contribute to other alterations, including the conversion of plants and animals to fossil fuels. Some subaerial oxidation processes convert ore minerals into more usable forms, others change them into less usable forms. Metals dissolved within the oxidation zone may be transported for reprecipitation below the water table. Metals removed from above thus serve to motivate a supergene enrichment within the upper part of the sulfide zone.

Metamorphism is a cyclical or episodic process that has served to change, during the past 4.6 b. y., the form and texture of minerals, including the creation and destruction of economic deposits. Changes in temperature and pressure, or in the quantity of water present, serve to stabilize some minerals, to unstabilize others. Typical results are recombinations and recrystallizations.

Barriers within a conduit tend to influence mineral deposition. These barriers may be of many types. Gougy, impervious, pre-mineral cross-fractures may precipitate cross-breaking seams. Wherever breccia-forming walls grade into gouge-forming walls, the change in the fault-filling coinciding with the contact may form an equally effective barrier and thereby confine the ore-bearing solutions to more competent layers, e. g., when a vein passes from quartzite to shale.

The greatest concentrations of ore are usually just below gently dipping barriers, except that, wherever the solutions have had a marked horizontal component, the ore concentrates on either the hanging-wall or foot-wall side of the barrier. The actual angle deflecting the flow within the barrier determines whether the shoot "bottoms" or "tops" there.

These and other factors relating to the flow of heat and material through the tectonosphere must be included as constraints in any backward extrapolation of the effects of intrusive and extrusive activity on the origin, evolution, and present characteristics of mineral deposits.

Other analyses related to the origin, evolution, and present characteristics of the paths of preferential flow within the Earth's tectonosphere include those by Bostrom et al. (1972), Dandurand et al. (1972), Fein (1973), Gill (1973), Griggs (1972), Hackett and Bischoff (1973), Heirtzler (1972), Murray (1970), Nitsan (1973), Roy et al. (1972), Steinhorn (1973), and Zonenshain (1972).

The Relationship between Intrusive and Extrusive Magma and the Low-Velocity Zone as Interpreted by the Tectonospheric Earth Model.

The low-velocity zone is related to intrusive and extrusive activity, according to the Tectonospheric Earth Model, through a common mechanism, i. e., the same driving mechanism motivates (1) magmatic activity and (2) the low-velocity zone, possibly produced by incipient melting in the upper mantle (See, e. g., Lambert and Wyllie, 1970). The postulated incipient melting is caused by the same mechanism that melts the magma that ascends to produce intrusive and extrusive activity. Thus, the low-velocity zone may be considered a complex global network, or system, of horizontal "intrusions" between asthenospheric plates and shells (Tatsch, 1972a: chapter 6). In these regions, the unbalanced force vector on some protomagma is horizontal (i. e., this may occur wherever there are insufficient vertical conduits to convey all the protomagma upward). Magma unable to proceed upward will proceed sideward, if there is a horizontal path, or it will remain in situ, if there are no paths. Intrusive and extrusive activity is closely related, therefore, to both the low-velocity zone and to the paths of preferential flow.

Other studies in related areas have been made by Dimitriyev (1972), Heirtzler (1972), Scheidegger (1972), Shtreys and Tseysler (1972), York and Helmberger (1973), and Zaplotal (1960).

The Association between Intrusive and Extrusive Magmatism and Tectonic Activity as Interpreted by the Tectonospheric Earth Model.

Basaltic and rhyolitic magmatism is associated with crustal thinning which, in turn, may be caused by (1) crustal tension, (2) heating from below, or (3) a combination of these. Because the amount of crustal tension and heating in a given area at a given time is dependent upon contributions from the geometrical, mechanical, thermal, and chemical factors of the subsurface environment, it may be said that intrusive and extrusive magmatism forms a temporal spectrum, or continuum, of compositions ranging from ultra-basicity to ultra-acidity. This compositional continuum repeats in a cyclical but aperiodic manner, according to the Tectonospheric Earth Model. This accounts, according to this concept, for the great variety of basic-to-acid rocks that have been emplaced on the surface of the Earth during the past 4.6 billion years (b. y.) (Tatsch, 1972a).

Other analyses in related area include those by Azhgirei (1968), Bain (1968), Benerjee and Ghosh (1972), Bickford and Van Schmus (1973), Borrello (1972), Cepeda (1973), Corliss (1973), Dixon and Pereira (1973), Gilluly (1971, 1972), Harrison (1972), Heirtzler (1972), Hutchinson and Hodder (1972), Kaula (1972), Khain (1972), McElhinny (1972), McGetchin et al. (1973), McKenzie (1972), Miyashiro (1972, 1973), Sillitoe (1972a, 1972b, 1972c, 1973), Spall (1972), Subhotin et al. (1972), Winterer (1973), and Wright and McCurry (1973).

Intrusive and Extrusive Activity Associated with Uplifts and Thrusts as Interpreted by the Tectonospheric Earth Model.

Many mountain ranges are the product primarily of vertical uplift, rather than of horizontal crustal compression as was once thought (See, e. g., Eardley, 1963). The Laramide Rockies, for example, east of the Paleozoic miogeosyncline, were formed primarily by intrusive activity on a large scale consisting essentially of oval or irregularly broad shapes having structural relief of a few hundred meters (e. g., the Bowdoin Dome) to 12 km (e. g., the Wind River Uplift). Later Tertiary faulting modified considerably some of these Laramide "intrusives". Younger Laramides were modified by erosion, sedimentation, and subsequent intrusive and extrusive activity. When the thrust faults of the Laramide province are charted, they prove to be, for the most part, marginal to the uplifts or intrusives. There is general correlation between the amount of uplift and the type and prominence of border thrusts, suggesting a common causal mechanism. This association is very pronounced in cases of large uplifts (i. e., 6 km or more) but they may be almost undetectable in intrusives of low and intermediate uplift.

These and similar relationships between uplifts and thrusts suggest that vertical uplift (or intrusion from below) was the primary deformation and that thrusting was a secondary lateral deformation caused by unbalanced force vectors acting vertically and obliquely downward, i. e., such as would be evidenced by gravity sliding, flowing, and large-scale erosion, for example (Tatsch, 1972a).

Many mineral deposits appear to be associated with (1) the volcanism of intermediate and silicic types related to uplifts and graben faulting; and (2) the flood-basalt extrusions related to massive crustal warping along future rift margins (See, e. g., Baker et al., 1972).

Kimberlites are usually associated with intrusive carbonate (carbonatite), alkali basalts (lamprophyres), or other unusual highly alkalic rocks in regions commonly associated with epeirogenic uplift (Wyllie, 1967, 1969a, 1969b). The genesis of these rocks appears to involve the localization of volatiles or partial melting in the upper mantle. Because this genesis apparently reflects long-lived, deep-seated mantle behavior, the spatio-temporal relationships of kimberlite petrology and mineralogy is of interest to the origin, evolution, and accumulation of other mantle-derived minerals.

Very few proposals exist for the origin of the volatiles that appear to be involved not only in kimberlite emplacement but also in epeirogenic uplift (See, e. g., McGetchin et al., 1973; Tatsch, 1972a).

Related analyses include some by Damon (1972), Faure (1972), Hills (1972), and Trofimov (1972).

Subduction, Obduction, and Other Omniductive Activity as Contributions to the Evolution of Mineral Deposits during the Past 4.6 Billion Years.

Many minerals seem to arrive at the Earth's surface as a result of sea-floor spreading, crustal plate motion, continental drift, and other omniductive movements involving lithospheric plates and the magmas associated therewith. It is well to consider briefly how these geophenomena are interpreted as constituting essential phenomena in the evolution of mineral deposits.

When viewed on a global scale and over a long period of time, sea-floor spreading and other forms of crustal rifting appear to represent the surficial behavior of the Earth in those areas where the Earth's internal behavior causes crustal tension. Crustal plate motion, subsurface plate and block motion, and continental drift, in a similar manner, represent the surficial behavior of the Earth in those areas where the resultant of the Earth's internal behavior is representable as a tangential unbalanced force vector acting on a plate, on a block, or on a continent to produce motion of that block, plate, or continent relative to the bulk of the Earth or to a datum therein (Tatsch, 1972a).

In short, sea-floor spreading, crustal plate motion, subsurface plate and block motion, and continental drift, according to the Tectonospheric Earth Model, are surficial adjustments to compensate for the disequilibrated and ever-changing behavior of the Earth's interior. The Earth's surficial behavior is a temporarily delayed attempt of the Earth's surface to adjust itself to the ceaseless equilibrating behavior of its interior. In some cases a block may be moved to effect equilibration; in others, a plate may be moved. In still other cases, a rift may be formed to effect the equilibration. Sometimes, worldwide mountain ranges grow from these rifts in a single orogeny. At other times, an earthquake occurs when a block or plate adjusts itself to a more nearly equilibrated position in the never-ending attempt of the Earth to assume a condition of minimum energy. These and related matters regarding the Earth's behavior are discussed in greater detail in Tatsch (1972a).

The Evolution of the Mineral Deposits during the Past 4.6 Billion Years according to the Tectonospheric Earth Model.

When the Tectonospheric Earth Model is used to analyze the evolution of the mineral deposits during the past 4.6 b. y., it is convenient, because of the inherent geometrical, mechanical, thermal, and chemical constraints of the model, to consider three aspects of the evolution of the mineral deposits: (1) their most probable evolutionary behavior during the period 4.6 b. y. to 3.6 b. y. ago; (2) their most probable evolutionary behavior during the period 3.6 b. y. to 0.6 b. y. ago; and (3) their most probable evolutionary behavior during the period 0.6 b. y. ago to the present. This is done in the following sections.

The Evolution of the Mineral Deposits during the Period 4.6 to 3.6 Billion Years Ago.

For the purpose of this analysis, it is assumed, for reasons stated above, that the Earth's surface 4.6 b. y. ago was devoid of any surficial features other than those that would have been genetic to an accretionary process of planetary formation from the solar nebula at that time (Tatsch, 1972a: chapter 1). The evolutionary behavior of the mineral deposits during the period 4.6 b. y. to 3.6 b. y. ago would have been motivated by the Earth's internal behavior plus the influence of extraterrestrial factors operating upon the Earth during that period.

According to the Tectonospheric Earth Model, the fragments of the five "terrestrial" octants of Earth Prime became accreted onto a 5400-km primordial Earth during the period 4.6 b. y. to 3.6 b. y.

ago. Present evidence does not define clearly exactly when this accretion occurred within this one-billion-year span. Nor is it necessary to this discussion that this be known. Suffice it here to say merely that the evolution of the mineral deposits during the period 4.6 b. y. to 3.6 b. y. ago depended primarily upon two basic aspects of the Earth's behavior during that period: (1) the mineral "mix" within the accreting fragments of Earth Prime was not as homogeneous as was that of the surface of the 5400-km primordial Earth; and (2) the primordial Earth, as it accreted to its present size, behaved in accordance with the geometrical, mechanical, thermal, and chemical aspects described in Chapters 2 and 3.

It is of interest to recall, in this connection, that many of these accreting fragments bore a marked resemblance to the rocks that now comprise the Moon, the asteroids, and the meteorites (Tatsch, 1972a: chapter 19; 1974a).

The above-described behavior produced the following constraints upon the evolutionary behavior of the mineral deposits during the period 4.6 b. y. to 3.6 b. y. ago: (1) the surface of the 5400-km primordial Earth was sialic material cut by three mutually-orthogonal belts of extruding mafic and ultramafic material; and (2) the accreting material (from the fragments of the five terrestrial octants of Earth Prime) comprised a complete spectrum of material, ranging from (a) fragments of iron, nickel-iron, and other metals that might have been contained within the molten and plastic parts of Earth Prime prior to its fragmentation, to (b) sialic material (from the surface of Earth Prime) that was similar to that of the surface of the 5400-km primordial Earth.

These constraints produced the following evolutionary behavior within the Earth's surface during the period 4.6 b. y. to 3.6 b. y. ago: (1) the initial sialic minerals on the surface of the primordial 5400-km Earth and on the erstwhile surface of Earth Prime were thoroughly metamorphosed by the effects of the accretionary activity and of the geometrical, mechanical, thermal, and chemical behavior of the primordial Earth as it grew from about 5400 km to its present size; and (2) as the size of the primordial Earth grew (at an average rate of about one millimeter per year), the geometrical, mechanical, thermal, and chemical effects of the Earth's internal behavior were felt mainly through the primordial orogenic belts of the Earth during the period 4.6 b. y. to 3.6 b. y. ago, i. e., through the wedge-belts of activity of the Tectonospheric Earth Model.

This evolutionary behavior within the Earth's surface (as it grew at an average rate of about 3 micrometers per day) produced the following effects upon the distribution patterns of the mineral deposits during that period: (1) the igneous, metamorphic, and sedimentary processes associated with the mineral deposits tended to form three mutually-orthogonal belts upon the surface of the Earth; and (2), as the Earth grew, at a rate of a few micrometers per day, it tended to continually equilibrate its condition to one of minimum energy by modifying the upper few micrometers of its surface each day, through the geometrical, mechanical, thermal, and chemical aspects of its driving mechanism (Chapters 2 and 3), thereby redistributing, metamorphosing, and otherwise affecting the mineral deposits to the extent that would be expected as a result of the geometrical, mechanical, thermal, and chemical effects of the basic driving mechanism under these conditions.

The most probable distribution pattern of the mineral deposits at the end of this period, i. e., 3.6 b. y. ago, may be summarized: (1) the youngest mineral belts (i. e., those 3.6 b. y. old on the present time scale) were aligned along various segments of three mutually-orthogonal belts; and (2) the older mineral belts consisted of "fossil" belts that had been overprinted, metamorphosed, and otherwise modified to the extent that would be expected as a result of the geometrical, mechanical, thermal, and chemical effects of the Earth's internal behavior during the period 4.6 b. y. to 3.6 b. y. ago.

The Evolution of the Mineral Deposits during the Period 3.6 to 0.6 Billion Years Ago.

For the purpose of this analysis, it is assumed, for reasons previously stated, that the Earth's surface 3.6 b. y. ago was that which would be expected from the evolutionary behavior described in the previous section. It is assumed further that, because most extra-terrestrially induced effects upon the Earth ceased prior to 3.6 b. y. ago, the Earth's behavior during the past 3.6 b. y. was internally motivated (Tatsch, 1972a). The evolutionary behavior of the mineral deposits during the period 3.6 b. y. to 0.6 b. y. ago would have been motivated almost entirely by the Earth's internal behavior during that period.

This behavior produced the following constraints upon the evolutionary behavior of the mineral deposits during the period 3.6 b. y. to 0.6 b. y. ago: (1) the surface of the Earth 3.6 b. y. ago consisted of three mutually-orthogonal orogenic belts plus modified older "fossils" as described in the previous section; and (2) new activity occurred primarily within the wedge-belts of activity associated with the Earth during the period 3.6 b. y. to 0.6 b. y. ago.

These constraints produced the following evolutionary behavior within the Earth's surface during the period 3.6 b. y. to 0.6 b. y. ago: (1) the younger mineral belts (i. e., those that are now 3.6 b. y. old) were metamorphosed by the effects of the geometrical, mechanical, thermal, and chemical behavior of the Earth during this period; and (2) some of the older "fossil" belts (i. e., those older than 3.6 b. y.) were modified by uplift, erosion, metamorphism, overprinting, and by other effects expected from the geometrical, mechanical, thermal, and chemical behavior of the Earth during that period.

This evolutionary behavior within the Earth's surface produced the following effects upon the distribution patterns of the mineral deposits during the period 3.6 b. y. to 0.6 b. y. ago: (1) the igneous, metamorphic, and sedimentary processes associated with the mineral deposits tended to form three mutually-orthogonal belts upon the surface of the Earth at any time during that period; and (2) the Earth tended to continually equilibrate its condition to one of minimum energy through the geometrical, mechanical, thermal, and chemical aspects of its driving mechanism (Chapters 2 and 3), thereby redistributing, metamorphosing, and otherwise affecting the mineral deposits to the extent that would be expected from the geometrical, mechanical, thermal, and chemical effects of the Earth's basic driving mechanism under these conditions.

The most probable distribution pattern of the mineral deposits at the end of this period, i. e., 0.6 b. y. ago, may be summarized: (1) the youngest mineral belts (i. e., those of the Upper Precambrian) were aligned along various segments of three mutually-orthogonal belts; and (2) the older mineral belts consisted of "fossil" belts that had been overprinted, metamorphosed, and otherwise modified as would be expected to result from the geometrical, mechanical, thermal, and chemical effects of the Earth's internal behavior during the period 3.6 b. y. to 0.6 b. y. ago.

The Evolution of the Mineral Deposits during the Period 0.6 Billion Years Ago to the Present.

For the purpose of this analysis, it is assumed, for reasons previously stated, that the Earth's surface at the end of the Precambrian was that which would be expected from the evolutionary behavior described in the previous section. It is assumed further that the Earth's behavior during the Phanerozoic was basically internally motivated (Tatsch, 1972a). The evolutionary behavior of the mineral deposits during the Phanerozoic have been motivated, under this concept, almost entirely by the Earth's internal behavior during this period.

This behavior produced the following constraints upon the evolutionary behavior of the mineral deposits during the Phanerozoic: (1) the surface of the Earth at the end of the Precambrian consisted of three mutually-orthogonal orogenic belts plus modified older "fossils" as described in the previous section; and (2) new activity has occurred primarily within the wedge-belts of activity associated with the Earth during the Phanerozoic.

These constraints have produced the following evolutionary behavior within the Earth's surface during the Phanerozoic: (1) the younger mineral belts (i. e., those of latest Precambrian age) were metamorphosed by the effects of the geometrical, mechanical, thermal, and chemical behavior of the Earth during the Phanerozoic; and (2) some of the older "fossil" belts (Precambrian) were modified by uplift, erosion, metamorphism, overprinting, and by such other effects that might be expected from the geometrical, mechanical, thermal, and chemical behavior of the Earth during this period.

This evolutionary behavior within the Earth's surface has produced the following effects upon the distribution patterns of the mineral deposits during the Phanerozoic: (1) the igneous, metamorphic, and sedimentary processes associated with the mineral deposits have tended to form three mutually-orthogonal belts upon the surface of the Earth at any time during the Phanerozoic; and (2) the Earth has tended to continually equilibrate its condition to one of minimum energy through the geometrical, mechanical, thermal, and chemical aspects of its driving mechanism (See Chapters 2 and 3), thereby redistributing, metamorphosing, and otherwise affecting the mineral deposits to the extent that would be expected from the geometrical, mechanical, thermal, and chemical effects of the Earth's basic driving mechanism under these conditions.

The most probable distribution pattern of the mineral deposits at the present time, according to this concept, may be summarized: (1) the youngest mineral belts (for example, the Tertiary copper belts and the Tertiary petroleum belts) should be aligned along various segments of three mutually-orthogonal belts; and (2) the older mineral belts consist of "fossil" belts that have been overprinted, metamorphosed, and otherwise modified as would be expected to result from the geometrical, mechanical, thermal, and chemical aspects of the Earth's internal behavior during the Phanerozoic.

Chapter 6

CENOZOIC SEISMOTECTONOMAGMATIC BELTS AND THE ASSOCIATED URANIUM DEPOSITS

The present distribution patterns of the Earth's uranium deposits are identifiable, according to the seismotectonomagmatic-belt concept, with the surficial manifestations of the Earth's internal behavior associated with the 12 quadrispherical arcs of the Tectonospheric Earth Model. The present configuration of these 12 arcs may be listed, together with some of the applicable geographical areas:

a. Bengal-to-Kermadecs. Australia, Bangladesh, Brunei, Burma, Indonesia (including West Irian, Java, Kalimantan, and Sumatra), the Khmer Republic, Malaysia, Papua New Guinea, southern Philippines, South Viet Nam, Sri Lanka, Thailand, and the associated offshore and sea-bottom areas.

b. Kermadecs-to-Galapagos. Islands plus associated offshore and sea-bottom areas.

c. Galapagos-to-Gibraltar. Central America, northern South America, northwestern Africa, Spain, and the associated offshore and sea-bottom areas.

d. Gibraltar-to-Bengal. Spain, northern Africa, southern Eurasia, the Middle East, Iran, Pakistan, India, and the associated offshore and sea-bottom areas.

e. Bengal-to-Aleutians. Indonesia, Malaysia, southeastern Asia, northeastern Asia, and the associated offshore and sea-bottom areas, including the South China Sea, the East China Sea, the Sea of Japan, the Sea of Okhotsk, and the Bering Sea.

f. Aleutians-to-Galapagos. Southern Alaska, western Canada, western United States, western Mexico, Central America, and the offshore and sea-bottom areas associated with this arc.

g. Galapagos-to-Bouvet. Western South America, eastern Argentina, and the associated offshore and sea-bottom areas.

h. Bouvet-to-Bengal. Southern Africa, southeastern Africa, southern India, southeastern India, and the associated offshore and sea-bottom areas.

i. Kermadecs-to-Aleutians. Southern Alaska, various islands between Alaska and the Tonga-Kermadecs complex, and the associated offshore and sea-bottom areas.

j. Aleutians-to-Gibraltar. Western and northern Alaska, Canadian Arctic islands, western and southern Greenland, western Europe, and the associated offshore and sea-bottom areas.

k. Gibraltar-to-Bouvet. Western Africa and the associated offshore and sea-bottom areas.

l. Bouvet-to-Kermadecs. Antarctica, New Zealand, the Tonga-Kermadecs complex, the associated islands, and the associated offshore and sea-bottom areas.

In considering these listings, it is suggested that the reader use a globe with 3 rubber bands to represent the surficial traces of the 12 quadrispherical arcs of the model. When properly placed on the globe, the 3 rubber bands will form 6 orthogonal intersections (i. e., quadruple points) associated with the 6 points of the model: Aleutians, Galapagos, Gibraltar, Bengal, Kermadecs, and Bouvet (Chapter 2). In doing this, it is well to recall (from Chapter 2) that the rubber bands represent lines that are roughly 1000 km beneath the surface of the Earth, whereas the known uranium deposits are at (or near) the surface. Because the projections of heat, volatiles, and other forms of energy through 1000 km of heterogeneous material, over a period of 4.6 b. y., is not a straight-line function (See Chapter 2), the projections must be made realistically, i. e., in terms of the Earth's geometrical, mechanical, thermal, and chemical heterogeneities (See, e. g., Tatsch, 1972a).

It is necessary to recall that the above listings, insofar as they pertain to the origin, evolution, and present characteristics of uranium deposits, are based on the first-order outputs of the Tectonospheric Earth Model. These outputs have a computed nominal probable error of 1571 km (Tatsch, 1972a: chapter 6). Resolutions smaller than 1571 km are not intended when making first-order analyses such as this. The resolutions obtainable with this model when using its higher-order outputs are one, two, and three orders of magnitude better (i. e., 157.1 km, 15.71 km, and 1.571 km), depending upon the amount of time and funding allocated to a given areas

No adjustments have been made in the above listings to correct the first-order predictions of the model for the differential horizontal movements that have occurred, between the Earth's present surficaal features and the subtectonospheric fracture system of the model (Tatsch, 1972a: chapter 6), since the time that the uranium deposits were emplaced in their present locations.

In the higher-order analyses, these adjustments, amounting to translations of as much as 140 km (and appropriate rotations) for some Cenozoic uranium deposits, are made for the regions being analyzed. More-complex adjustments must be made in analyses that involve "older" uranium deposits. These adjustments may amount to 300 km for Mesozoic deposits (Chapter 7) and 500 km for Paleozoic deposits (Chapter 8).

The important point regarding these adjustments is that the Tectonospheric Earth Model provides a method for determining them to any degree of accuracy, limited only by the amount of time and funding available. Also, because of the "limit stop" aspects of the orogenic-cratonic evolution of the continents (Tatsch, 1972a: chapter 7), the <u>sum</u> of these adjustments rarely exeeeds 3500 km, except that adjustments for some of the oldest Archean uranium deposits could amount to as much as 4900 km under the seismotectonomagmatic-belt concept of the Tectonospheric Earth Model. No <u>single</u> adjustment under this concept amounts to more than a few km, however, because individual adjustments are computed and applied incrementally in the entire gamut of the spatio-temporal frame spanning 4.6 b. y. to a depth of 1000 km (Tatsch, 1973a).

<u>Representative Cenozoic Uranium Deposits</u>.

In order to determine how well the Earth's actual uranium deposits follow the patterns predicted by the Tectonospheric Earth Model, it is well to consider the locations and characteristics of representative examples of today's known Cenozoic uranium deposits.

<u>Agostinho</u> (Brazil).

This deposit, part of the Pocos de Caldas (q. v.), contains uranium as an amorphous mineral disseminated within the tinguaitic breccia and as small quantities of urano-thorianite and coffinite. The uranium is associated with fluorite, pyrite, and the minerals of thorium and molybdenum. The uranium grades about 0.2% U_3O_8, mostly in brecciated and subvertical veins (a few meters thick) that cut the tinguaitic rocks near their contact with fayaites.

This deposit resembles Cercado (q. v.).

Algeria (Africa).

This deposit is like Morocco (q. v.).

Angola (Africa).

A broad belt of late Mesozoic and Cenozoic uraniferous phosphate rocks stretches from Cabinda to southwestern Angola. These low-grade deposits range from about 0.05% to 0.2% U_3O_8 (See, e. g., Bowie, 1970). Most of the mineralizations are lenticular.

Arlit (Niger).

These peneconcordant uraniferous ores in standstone contain 26,000 tons of ore, averaging 0.29% U_3O_8. See also Niger.

Ascension (Colorado).

See Golden Gate Canyon.

Austrian Alps (Europe).

Some of the Permo-Triassic sandstone uranium deposits of the Austrian Alps were mobilized by Alpine seismotectonomagmatic-belt activity (Tatsch, 1973a). See also Chapter 8.

Baghal Chur (Pakistan).

The Baghal Chur deposit of the Dera Ghazi Khan area lies within the Middle Siwalik formation (middle Miocene to late Pliocene) franking the Sulaiman Range. The mineralization and the associated topography are vestiges of seismotectonomagmatic-belt activity identifiable with the Aleutians-Bengal-Bouvet hemispheric wedge-belt of activity (Tatsch, 1973a), striking orthogonally to the Gibraltar-Bengal-Kermadecs hemispheric wedge-belt of activity of the Cenozoic.

Uraniferous lenses are known within seven localities extending over a strike length of almost 200 km along the foothills of the Sulaiman Range (See, e. g., Moghal, 1974). The best of these are at Baghal Chur. The unraniferous manerals are metatyuyamunite (oxidized) and uraninite-coffinite (unoxidized).

The mineralized lenses range up to 100 m in length and 3 m in thickness. The denudation of the rising Himalayas appears to have provided the uranium. Controls and transport appear to have been faciliteted by the vestiges of seismotectonomagmatic-belt activity found in that area, particularly those associated with the Aleutians-Bengal-Bouvet wedge-belt of activity. The uranium content ranges from 0.05% to over 0.5% U_3O_8.

The paleochannels associated with the conglomerate lenses are reminiscent of those of western U. S. A., including the Wyoming Basin and the Colorado Plateau (q. v.). The Baghdal Chur ore is light gray, mostly medium-to-fine grained, soft, friable, and poorly sorted. Some pebbles of quartzite, limestone, and cal careous clay also occur. The major heavy mineral is magnetite.

Two genetic concepts exist for the uranium: (1) clastic heavy mineral particles; and (2) ground-water sediments similar to those of western U. S. A. Other details of these deposits are contained in the literature (See, e. g., Basham and Rice, 1974; Moghal, 1974).

Balger (Saskatchewan).

See Middle Lake.

Bendada (Portugal).

See Serra da Estrada.

Benevides (Texas).

The Benevides deposit, on the Gulf Coastal Plain, is a roll-type deposit. This deposit supports a bi-partite division of the altered tongue. It also contains a mineral suite that limits the possible explanations for the geochemical processes that produce the ore and the alteration.

As much as half a km ahead of the roll front, ulaltered rock contains a trace of magnetite, about 0.1% ilmenite, and about 1.0% pyrite. Just back of the roll front, the altered rock is pale orange, the ilmenite contant has dropped to about 0.05%, the magnetite (partly altered to hematite) content is roughly 0.03%, goethite comprises about 0.15% of the rock, and pyrite is absent. Roughly 1 km behind the roll front, the altered rock is pale red, contains about 0.15% authigenic magnetite (partly altered to hematite), contains about 0.3% ilmenite, and goethite and pyrite are practically zero. Clay fractions are very red in the hematitic zone, yellow in the alteration envelope, and pale gray in the unaltered rock.

Bone Valley (Florida).

The Pliocene phosphorite in the Bone Valley Formation is about 3m thick and covers an area of roughly 500 km^2. It averages 0.012% to 0.024% U_3O_8 and 20% to 30% P_2O_5.

Bruni (Texas).

This deposit, about 65 km east of Laredo, is being mined by in-situ leaching (See, e. g., Crawford, 1975). This operation will be similar to that at Clay West (q. v.) and other in-situ operations along the Texas Coastal Plain (q. v.). Like Clay West, this deposit lies along a vestige of seismotectonomagmatic-belt activity that stretches at least from Mexico to Louisiana (Tatsch, 1973a).

Buller Gorge (New Zealand).

The uranium in this deposit occurs in at least 10 horizons in sandstones of the Hawks Crag breccia. The grade is only about 0.05% U_3O_8. These deposits resemble those of the Colorado Plateau (q. v.), both in mode of occurrence and in the presence of coffinite and pitchblende as the main uraniferous minerals.

Calaf (Spain).

These deposits are similar to those in Palencia (q. v.).

Clay West (Texas).

This deposit, about 15 km southwest of George West, is being mined commercially as an in-situ leaching operation. Comprising 66 injection wells and 46 extraction wells, Clay West extracts about 250,000 lb/yr of U_3O_8 from sandstones at depths to 170 m (See, e. g., White, 1975). The unaniferous province being tapped lies along a vestige of seismotectonomagmatic-belt activity that stretches from north of Houston to Brownsville (Tatsch, 1973a). Other in-situ leaching projects are planned along this vestige (See, e. g., Crawford, 1975). The uranium ranges from 0.05% to 0.5% U_3O_8 in Oakville sandstone. The Oakville is underlain by Catahoula clays, which have an abnormal U_3O_8 content, apparently derived from volcanic ash deposits during the Catahoula episode of seismotectonomagmatic-belt activity. Percolating ground waters dissolve the U_3O_8 and transport it to nearby rocks. There, in the presence of H_2S from petroliferous formations, the U_3O_8 precipitates (Tatsch, 1974b).

Other details of this deposit are described in the literature (See, e. g., White, 1975; Crawford, 1975). See also Texas Coastal Plain.

Crooks Gap (Wyoming).

This deposit is just south of and similar to Gas Hills (q. v.). See also Fischer (1974b).

Duval County (Texas).

This deposit is similar to Felder (q. v.).

Edmund (Western Australia).

See Mundong Well.

Egypt (Africa).

This deposit is like Morocco (q. v.).

Fakili (Turkey).

The Fakili deposit, in the Aegean region of western Turkey, contains an average concentration of 0.044% U_3O_8 (See, e. g., Kaplan et al., 1974). The rocks of that area are mostly of the metamorphic series of the Menderes Massif and Neogene sediments. Augen-gneisses and other metamorphic rocks comprise the basement of the Neogene sedimentary basin. The Fakili deposit occurs within a facies characterized by: (1) variegared coloring; (2) the absence of marl; and (3) the presence of secondary gypsum and sedimentary pyrites. The deposit was formed in two stages: (1) syngenetic precipitation, and (2) epigenetic concentration. Each of these stages is identifiable with the preferential paths of fluid flow associated with seismotectonomagmatic-belt activity (Tatsch, 1973a).

Felder (Texas).

The Felder deposit, in Live Oak County of the Texas Coastal Plain (q. v.), is a roll-type deposit. In may respects, this is a typical roll-type deposit, except that (1) the rock in the altered tongue contains pyrite, and (2) there is about as much pyrite in the unaltered rock away from the ore as there is within the ore itself. Magnetite, sparse throughlut the pyrite, is more abundant in some of the oxidized rock. Local redistribution of ore from the original deposit occurred wherever weathering affected the ore, as evidenced by irregularities in ore distribution and radioactive equilibrium. There is a possibility that the altered tongue once had a typical bleached alteration envelope and a hematitic core, but recent weathering has obliterated these in many places.

The most probable source for the uranium is volcanic ash, which is an integral component of the host rocks. Many of these rocks are related to normal faulting and associated vestiges of seismotectonomagmatic-belt activity. These vestiges trend parallel to the outcrop strike. The reductant appears to have been supplied through paths of preferential flow associated with the vestiges of seismotectonomagmatic-belt activity (Tatsch, 1973a). These paths provided the reducing fluids from underlying hydrocarbon reservoirs and other carbonaceous material.

Other details of this deposit are contained in the literature (See, e. g., Granger and Warren, 1974; Klohn and Pickins, 1974; Eargle et al., 1975). See also Texas Coastal Plain.

<u>Ferghana</u> (U. S. S. R.).

These carnotiferous sandstones are similar to those of the Colorado Plateau (q. v.). See also USSR Black Shales in Chapter 9.

<u>Gas Hills</u> (Wyoming).

The uranium deposit of the Gas Hills district is of the roll type. Nearly all the altered rock of the district is very pale gray to nearly white. A transition zone separates the altered tongue from the ore. One of the primary differences between the altered and unaltered rock is in the shapes and sizes of the pyrite grains: (1) in the altered tongue, there are no limonite specks, and the sparse pyrite is very fine-grained, shiny, and euhedral; and (2) in the ore zone, there are scattered limonite specks, and the pyrite is even finer-grained and occurs in clusters that are less shiny than those of the altered zone. The pyrite clusters are tarnished and alter quickly after exposure to the air. Unoxidized rock away from the ore contains untarnished coarser-grained pyrite crystals associated with some magnetite.

Other details of this deposit are contained in the literature (See, e. g., Armstrong, 1974a; Fischer, 1974b; Dooley et al., 1974; Seeland, 1975).

Golden Gate Canyon (Colorado).

Pitchblende and secondary uranium mineralizations are found in and near vestiges of the Laramide episode of seismotectonomagmatic-belt activity. These vestiges, mainly fault zones, cut gneiss, pegmatite, and schist of Precambrian age. The fault zones, locally called "breccia reefs", are many meters wide and many km long. These reefs comprise ankerite and potash feldspar as a matrix enclosing partly replaced rock fragments. The pitchblende and base-metal sulfides occur as thin films and colloform masses coating ankerite crystals. Irregularly distributed through intensely altered wall rock, some of these are replaced by later sulfides.

The mineralization appears to have been localized mainly by the composition and texture of the wall rock in conjunction with the oxidation of the iron from ferrous to ferric form. Pitchblende veins of the carbonate occur wherever ample ferrous iron is available in the wall rock. The average grade is about 0.23% U_3O_8. (Von Backstrom, 1974b).

Grants (New Mexico).

These deposits are late Mesozoic and early Cenozoic. They are discussed in Chapter 7.

Hamra (Spanish Sahara).

The uraniferous content of Hamra, the world's largest single phosphate deposit, is not known. An estimate of 0.015% U_3O_8 is probably fairly good, based on the contiguous Moroccan deposits.

Highland (Wyoming).

See Powder River.

Himalayan Foothills (India).

The fluvial and lacustrine sandstones, clays, and boulder conglomerates contain 0.0005% to 0.03% U_3O_8. Some vertebrae fossils grade as high as 0.02% U_3O_8.

Israel (Mid East).

Like Morocco (q. v.).

Italian Quaternayy (Italy).

Uraniferous iron-sulfide, exhalative supergene mineralizations are found in 3 volcanic districts of central Italy (See, e. g., Mittempergher, 1970). The mean uranium content of the volcanites is 0.0025%. This mineralization, which is related to magmatic H_2S exhalations and to supergene weathering of the exhalative pyrite and marcasite, follows a hydrothermal rather than stratigraphic horizon. Some uraniferous bodies in this area have been found near small streams.

Japanese Neogene Deposits (Japan).

Uranium deposits were found, in 1955, at Mingyotoge, in southern Honshu. Since then, at least 20 other significant deposits have been found on this island and on Hokkaido and Kyushu. These include deposits at Akatani, Futoro, Hanamaki, Iida, Jyoban, Kanamaru, Kuchiwa, Minawa, Mitoya, Nakajyo, Nakamuruke, Noto, Okushiri, Okutango, Oouchi, Sunagawa, Turumizu, Tozawako, and Tono. These deposits comprise various combinations of autunite, adsorbed uranium, coffinite, ningyoite, ranquillite, torbernite, uraninite, uranian apatite, and uranociricite. The accompanying minerals are primarily pyrite and some grauconite, and gypsum. The lithology is largely sandstone with some conglomerate, mudstone, pyroclastics, and shale. The sediments are largely continental with some deltaic and neritic facies. Carbonaceous matter is associated with almost all of the deposits, and the most important deposits are located in highly carbonaceous sediments. The basement is primarily granitic with some granodiorite, mainly of pre-Tertiary age. The deposits are generally flat, lenticular, and concondant. In a few of the deposits (e. g., Ningyotoge and Tono), ore shoots occur along the dikes, faults, and shear zones cutting the host sediments.

The uranium appears to have been supplied from surrounding source rocks and to have been transported by circulating groundwater solutions. Other details are contained in the literature (See,, e. g., Doi et al., 1975; Hayashi, 1965, 1970).

Jorinji (Japan).

This deposit of the Tono area (q. v.) includes an oxidized body containing uranociricite as the main ore mineral (See, e. g., Hayashi, 1965). It appears that barium was leached from the weathered feldspars of the basement granites. The resulting precipitate contained both barite and uranociricite. See also Tsukiyoshi, a similar depoist that is also in the Tono (q. v.) mine area.

Karnes County (Texas).

This deposit is similar to Felder (q. v.) except that this deposit is in a regressive near-shore marine sandstone facies of the late Eocene Jackson group, whereas the Felder deposit is in fluvial sandstones of the Miocene Oakville formation (See, e. g., Pickens, 1974).

Other details are contained in the literature (See, e. g., Adams and Weeks, 1974; Eargle et al., 1975; Brooks, 1975).

Specific mines include: Pfeil, Wright, McCrady, Weddington-Conoco, Weddington-Tenneco, Weddington-Susquehana, Butler, Pawelek, Galen, Bosco, Bargman, Hackney, Lauw, Nuhn, Luckett, Sickenuis, Brysch, Manka, and Stoeltje.

Kurayoshi (Japan).

These deposits and those at Ningyo-toge are associated with Paleozoic-Mesozoic sediments in the vicinity of the granitic Chugoko Massif. Similar deposits occur in the vicintiy of other Japanese granite massifs. The deposits, of the sandstone type, resemble those of the Colorado Plateau (q. v.), with coffinite and pitchblende as the main uraniferous minerals. The average grade is 0.05% U_3O_8 in conglomerate horizons.

La Bajada (New Mexico).

This deposit, also called Lone Star, is about 35 km southwest of Santa Fe. The urano-organic material is associated with sulfide mineralization along a fault in the Oligocene altered tuff-breccia. The deposits appear to be hydrothermal and to represent magmatic vestiges of seismotectonomagmatic-belt activity (Tatsch, 1973a). The uraniferous material is underlain by the coal and oil-bearing Mesa Verde units. The uranium is associated with submicroscopic mineral matter disseminated through the organic components.

La Coma (Mexico).

This deposit, in the State of Tamaulipas, is associated with uraniferous vestiges of seismotectonomagmatic-belt activity (Tatsch, 1973a). Estimated size is about 2000 tons.

Larap (Philippines).

This uraninite deposit, in Camarines Norte district, is of the replacement type. The goloogical setting comprises metamorphism, intermediate volcanics, basics, and ultrabasics. The ore minerals are chalcopyrite, galena, magnetite, molybdenum, pyrite, and uraninite (Santos, 1974). It is interesting that uraninite occurs within a chalcopyrite-molybdenum adjacent to a massive magnetite orebody (Tatsch, 1975g).

Libya (Africa).
Like Morocco (q. v.).

Live Oak County (Texas).
See Felder.

Lone Star (New Mexico).
See La Bajada.

Mabel New (Texas).
This is the only oxidized uranium deposit in that area of Texas. See Felder and Texas Coastal Plain.

Malargüe (Argentina).
This deposit, in the Province of Mendoza, is contained in sandstones and conglomerates (See, e. g., Stipanicic, 1970). These are associated with vestiges of late Mesozoic and early Cenozoic seismotectonomagmatic-belt activity (Tatsch, 1973a). In many respects, this deposit resembles Tonco-Amblayo (q. v.) farther north and probably associated with the same vestiges of seismotectonomagmatic-belt activity.

Middle Lake (Saskatchewan).
This supergene deposit is within the Athabasca region, which was above sea-level during much of the Phanerozoic. The resulting denudation caused peneplanation and the production of the supergene

alteration has extended to 50 m. The most common secondary uranium minerals are liebergite and uranophane. The average grade is 0.7% U_3O_8. Middle Lake is 16 km east of Stony Rapids. An associated mine is Balger, east of Verna Lake.

Misano (Japan).

See Tsukiyoshi.

Morocco (Africa).

These uraniferous marine phosphorites contain about 0.015% U_3O_8. The phosphate rocks comprise about 30 billion tons.

Mount Spokane (Washington).

Several areas on and near Mount Spokane contain small quantities of utunite and meta-autunite. The host rocks are alaskite, graphite granite, and biotite quartz monzonite. These are cut by numerous alaskitic and granitic pegmatite dikes and stringers. The ore minerals coat small joint, shear, and fracture surfaces. Some coarse crystals of meta-autunite fill fractures in shears and in open spaces within the rocks. Samarskite occurs within some pegmatites and as an accessory within alaskite. The host rocks appear to be vestiges of Laramide episodes of seismotectonomagmatic-belt activity (Tatsch, 1973a). Adjacent similar deposits include Quartz Ridge and Lost Creek in Pend Oreille County. This same uraniferous vestige extends northward into British Columbia. Part of the same vestige includes deposits within (1) the mountain range west of the Columbia and Kittle rivers, in Washington, and (2) north of Grand Forks, B. C. Some of these contain as much as 1% U_3O_8.

Mundong Well (Western Australia).

It is possible that pitchblende of Tertiary age is present at Mundong Well. But, if so, this is the result of reworking of the

Proterozoic uranium in this area. The deposit is described in Chapter 9.

Nacimiento-Jemez (New Mexico).

The deposits in this area, which span the Pennsylvanian-to-Tertiary age, are discussed under this heading in Chapter 7.

Niger (Africa).

In the case of the Niger mineralization, the sedimentological and paleo-geographic studies seem to indicate that the origin of the sediments constituting the host rock lies several hundred km to the south of the deposits (See, e. g., Gangloff in Moghal, 1974). Most deposits lie within ancient river beds. They resemble the deposits of the Colorado (q. v.) sandstones, both in general disposition and in the average grade of about 0.25% U_3O_8. Other details are contained in the literature (See, e. g., Bigotte and Obellianne, 1968).

Nigeria (Africa).

Some of these aluminum phosphates contain roughly 0.01% U_3O_8. Older (Mesozoic) uraniferous deposits in riebeckite granite also occur in Nigeria.

Ningyo-Toge (Japan).

These uraniferous ores in Neogene conglomerate contain about 4,000 tons of ore averaging roughly 0.06% U_3O_8. The mineralizations are in the basal part of the Neogene, immediately overlying a Mesozoic basement (See, e. g., Kamiyama et al., 1973).

Nuevo Leon (Mexico).

The State of Nuevo Leon contains many vestiges of seismotectono-magmatic-belt activity that appear to be uraniferous (Tatsch, 1973a).

Oliphants (Africa).

The west coast of southern Africa, entending roughly 1500 km northward from the mouth of the Oliphants River, through South West Africa, and into Angola, contains some uraniferous vestiges of seismotectonomagmatic-belt activity (Tatsch, 1973a). Within these vestiges, uranium-dissolving graoundwaters have served to reconcentrate roughly 100 deposits from uranium-rich rocks and veins that were eroded from earlier vestiges of seismotectonomagmatic-belt activity. Due to transgressions in that area, it is easily possible that oxidation was non-existent for a period of 35 m. y. during the Eocene and Oligocene (Von Backström, 1974b)

Palencia (Spain).

The Miocene sands of the province of Palencia contain dispersed organic matter showing strong radiometric anomalies. The same may be said of the Oligocene and Miocene lignites. These low-grade deposits are quite extensive and should become of economic interest after further exploration (See, e. g., Martin-Delgado-Tamayo and Fernandez-Polo, 1974). Similar deposits occur north of the Iberian Cordillera, including those at Calaf, Mequinenza, and Santa Coloma de Queralt.

Phosphoria (U. S. A.).

See Chapter 8.

Pinjor (India).

The vertebrate fossils of the Pinjor region contain uranium concentrations ranging from 0.05% to 0.34% U_3O_8 (See, e. g., Udas and Mahadevan, 1974).

Pocos de Caldas (Brazil).

The uraniferous alkalic, elliptical pipe or laccolith, in the southwestern part of Minas Gerais, crops out over an area of about 800 km^2. Like other volcanic features of this part of Brazil, this pipe is about 70 m. y. old. Activity within this vestige of a seimsotectonomagmatic-belt episode involved volcanism, intrusions of alkalic magma, at least two tectonic adjustments, and hydrothermal activity within the massif (Tatsch, 1973a). This episode affected also the Precambrian basement and the Mesozoic sediments of that area. Included within that area are the molybdo-uraniferous mineralizations of Agostinho (q. v.) and Cercado (q. v.).

The rocks are mostly syenites, phonolites, and associated tuffs. The in-situ deposits are within a net of thin veins; eluvial and alluvial mineralizations are scattered through a 450-km^2 drainage area. The uranium averages about 0.1% in association with (1) inclusions within zirconium minerals, and (2) in the crystal structure of the zircon. Most of the uranium is of the second type and therefore comprises an uneconomical refractory ore (Von Backström, 1974b).

Powder River (Wyoming).

Like most Tertiary basins of Wyoming, the Powder River Basin uranium deposits are excellent examples of large roll-type ore bodies. The host rocks are Paleocene sandstones deposited, as point bars, by a meandering stream. The source of the uranium appears to have been tuffaceous and arkosic debris indigenous to the sedimentary sequence containing the host rocks (See, e. g., Dahl and Hagmaier, 1974).

The largest deposits of best grade occur near the distal margins of permeable, slightly-dipping sandstones wherever they grade laterally into organic-rich siltstones, claystones, and lignites that were emplaced in backswamp or flood-basin environments. The deposits are epigenetic, apparently formed by precipitation of uranium from groundwater solutions moving through the host rocks from a recharge area southwest of the deposits toward a discharge area northeast of the deposits. This flow appears to have been northeasterly along vestiges of seismotectonomagmatic-belt activity in that area.

The deposits are large because the host rocks are extensive and the gound-water system remained fairly stable. Early pyrite formation appears to have been induced, in and around the host units, by a biogenic process using sulfate-reducing bacteria. Later oxidation of the pyrite, supposedly by the groundwater, caused sulfite to form. This, in turn, separated into the SO_4 and HS ions to develop the final reducing mechanism for precipitating the uranium in the ore rolls. Total tonnage is about 9 million tons, grading from 0.1% to 0.03% U_3O_8.

The Powder River Basin contains a variety of relationships between rock types and ore zones. If a generalization may be made, it is that the limonitic and bleached envelope is more extensively developed in the southern than in the northern parts of the red beds of the basin. An alteration envelope with variable characteristics borders parts of the hematitic cores of the altered tongues associated with the rolls of the basin. In the northern part of the basin, there are two color phases in the near-surface host rocks: (1) drab sandstones containing the uranium ores, and (2) hematitic red sandstones next to the ore (Sharp et al., 1954). In the southern part of the basin, goethite characterizes the rocks in contact with the uranium deposits, with the hematite in altered rock farther from the ore.

Some observations suggest that the sandstones were derived from a northern source, since the streams, in those days, flowed southeasterly (See, e. g., Law et al., 1975). Other details are contained in the literature (See, e. g., Granger, 1966; Fischer, 1974b).

Qatrani (Egypt).

The Qatrani deposit occurs within Oligocene sediments in the northern part of the Western Desert, roughly 100 km southwest of Cairo. The mineralization occurs within various forms, including phosphatic sandstone, ferruginated sandstone, carbonaceous clay, ordinary clay, limestone tubes, and fossil wood. Most of the uranium is very finely dispersed; the only uraniferous mineral known there is uraniferous francolite. Most of the mineralization occurs in bedded form as lenses, ranging from a few cm to a few m thick and tens of km long. The richest mineralization occurs along northwest-southeast and northeast-southwest tranding vestiges of seismotectonomagmatic-belt activity now represented by faults (Tatsch, 1973a).

In the uraniferous phosphatic sandstone, the uranium content ranges from 0.02% to 0.21%, except that within the francolite it may reach 0.6% (See, e. g., El-Shazly et al., 1974). In the uraniferous carbonaceous clay, the uranium content may reach 0.07%. Concentration normally average less in the other uraniferous deposits associated with Qatrani.

The Oligocene Gebel Qatrani Formation conformably overlies the upper Eocene Qasr El Sagha Formation. The Gebel consists of variegated sandstone, sands, and clays with recurring beds of impure limestone. A mid-Tertiary basalt flow occurs between the Oligocene sediments and the lower Miocene sediments in the Qatrani area. Tubes have been observed in the eastern part of the area in association with faults and in the sandy extension of the bone beds. The tubes range up to 20 cm in diameter and from 10 to 170 cm in height. The primary structures are the bedding and crossbedding of the sediments, apparently associated with vestiges of seismotectonomagmatic-belt activity. Secondary structures, also associated with these vestiges, include faults, folds, and joints. The older vestiges, trending northwest-southeast and northeast-southwest, appear to have been developed in conjunction with the land emergence during the upper Eocene. These vestiges are correlatable with the trends of the gulfs of Aqba and Suez and with the seismotectonomagmatic-belt activity that caused the Red Sea development (See, e. g., Tatsch, 1972a). Mineralization is present along many of the faults associated with these older vestiges, particularly as uraniferous ferruginous sandstone and uraniferous tubes. The younger vestiges trend east-west, westnorthwest-eastsoutheast, and eastnortheast-westsouthwest. The faults associated with these vestiges traverse all the rocks outcropping in the area, including the basalt sheets and the lower Miocene sediments. These vestiges are correlatable with the land subsidence associated with Mediterranean tectonism. A major anticline is superimposed on the Oligocene sediments with parallel minor anticlines and synclines.

The exogenic conditions that prevailed during uranium disposition were fluviatile to fluviomarine. This produced the observed sequence of sandstone and clay with occasional thin beds of impure limestone. Hot brines, of an acidic oxidized nature, are considered to have been the mineralizing solutions (See, e. g., El-Shazly et al., 1974). Although the source of the Red Sea hot brines is still controversial (See, e. g., Tatsch, 1975a; Ross, 1972), it appears that the uraniferous brines in the Qatrani area were derived from the volcanic or subvolcanic vestiges of seismotectonomagmatic-belt activity, as modified by the subsequent exogenic environment. These mineralizing solutions were probably introduced episodically rather than continuously (See, e. g., Tatsch, 1973a).

Ramshahr-Kalka-Morni (India).

Concentrations of uranium occur within a lower-to-middle Siwalik formation in this area. The most promising concentrations are in (1) hard gray to greenish sandstones, and (2) the interbedded intraformational clay conglomerates carrying unoxidized grayish mudstone, shale, and siltstone fragments plus carbonaceous and sulfide materials. Concentrations range from 0.02% to 0.6% U_3O_8. Uranophane is the most common uranium mineral.

Reboleiro (Portugal).

See Serra da Estrada.

Recife (Brazil).

The phosphorites of this area average 0.02% uranium. Otherwise, they resemble the deposits at Bone Valley (q. v.).

Ross-Adams (Alaska).

This deposit, of late Cretaceous or early Tertiary age, is described in Chapter 7.

Russia (Eurasia).

See U. S. S. R.

Sea Water (Oceans).

Sea water contains 0.00015% to 0.0016% uranium. Future production from sea water includes uranium byproducts from (1) desalination plants, and (2) magnesium-extraction plants.

Senegal (Africa).

Like Nigeria (q. v.).

Serra da Estrada (Portugal).

Four vein-type deposits occur within vestiges of Hercynian and Alpine episodes of seismotectonomagmatic-belt activity: Bendada, Guarda, Reboleiro, and Urgeirica. These are related to the Serrada da Estrada uplift created by the Alpine episode. The host granite is of Hercynian age with about 50 ppm (=0.005%) uranium. The veins follow a complex but generally northeast-southwest fracture system related to the Alpine episode of seismotectonomagmatic-belt activity (Tatsch, 1973a). The uraniferous fractures and the pitchblende are of Tertirry age. The three main types of veins are red jasper, ferruginous white quartz, and black and white banded quartz. Generally, the U_3O_8 content correlates with the total amount of sericitation involved. It appears that the uranium and iron originally present in the accessory minerals in the sericitized zones of the Hercynian granite were leached and mobilized during the Alpine episode of seismotectonomagmatic-belt activity and then concentrated and redeposited in more open sections of the fracture system.

Shirley Basin (Wyoming).

The Shirley Basin deposits lie within an intermontane-basin. Similar basins lie along the entire length of an Arctic-to-Argentina seismotectonomagmatic belt (See, e. g., Tatsch, 1973a; Buzzalini and Gloyn, 1972). These basins are identifiable with vestiges of Cenozoic seismotectonomagmatic-belt activity associated with the Aleutians-to-Galapagos and Galapagos-to-Bouvet quadrispheric arcs (Tatsch, 1973a).

The uranium deposits of the Shirley basin are of the roll type. Iron oxide minerals are most abundant in the altered rocks within 2 m of the ore; pyrite is virtually absent in the altered rock; and epidote and hornblende are least abundant in and near the ore zone. The altered tongues are mostly light yellowish-green as a result of a few % of iron-rich montmorillonite or nontronite, plus some goethite and hematite. There is no well-defined alteration envelope. The unaltered rock is very light gray and contains an iron-poor clay.

In may respects, the Shirley Basin deposits resemble those of the Texas Coastal Plain (q. v.). Other details are contained in the literature (See, e. g., Harshman, 1972; Warren, 1972; Fischer, 1974b Dooley et al., 1974; Ludwig, 1975).

<u>Sila</u> (Italy).

These mineralizations of autunite are on the Sila Plateau of Calabria. Their origin is unknown because the mineralizations do not display any obvious genetic associations with possible primary uranium occurrences in the reduced state (See, e. g., Dall'aglio, 1974).

<u>Spors Mountain</u> (Utah).

More than 40 uraniferous fluorspar deposits occur on Spors Mountain, Juab County. The rocks are mainly marine limestone, quartzite, dolomite, and related types, extending from the lower Ordovician to the middle Devonian. These rocks are cut by Tertiary plugs and dikes of dellenite, latite, and rhyolite. Some volcanic pipes filled with breccia occur on the eastern side of the mountain. The uranium content of the pipes varies widely between 0.003% and 0.33% and lenses out rapidly with depth. The fluorspar and uranium appear to have been produced from the latter stages of the same Tertiary seismotectonomagmatic-belt episode that produced the plugs and dikes. In some respects, the volcanic pipes resemble those in Arizona, Canada, Africa, and other parts of the world.

<u>Stony Rapids</u> (Saskatchewan).

See Middle Lake.

<u>Tamaulipas</u> (Mexico).

See La Coma.

<u>Texas Caliche</u> (Texas).

The caliche (calcrete) of Texas stretches from the western

Panhandle to the coast. This caliche is a vestige of the same type of seismotectonomagmatic-belt activity that produced the uraniferous caliche in Australia (Tatsch, 1973a). This suggests that the Texas caliche contains above-average concentrations of uranium. Some field observations have confirmed this (See, e. g., Finch, 1975).

Texas Cenozoic (Texas).

Most of the well-known Texas uranium deposits are in the Texas Coastal Plain (q. v.). But many less-well-known deposits occur in other parts of Texas. These are acsociated with vestiges of seismotectonomagmatic-belt activity stretching from the western Panhandle to the coast. Specific formations include the Ogallala (Miocene) and some limestone, marl, and volcanic ash of Pleistocene age in west Texas.

Texas Coastal Plain (Texas).

These uraniferous deposits are closely associated with the geosynclinal and faulted vestiges of late Paleozoic and Mesozoic seismotectonomagmatic-belt activity (Tatsch, 1973a). The hinge line of these vestiges is present in the general area of the uranium deposits (See, e. g., Eargle et al., 1975).

Some of these roll-type uraniferous peneconcordant deposits have been analyzed by Eargle et al. (1971), by Eargle and Weeks (1973), and by Eargle et al. (1975). The uraniferous tuffaceous sandy sedimentary rocks of the late Eocene to Pliocene also contain some molubdenum and selenium. Both yellow (lightly oxidized) and and black (unoxidized) ore occur, with the oxidized ore near the surface and the unoxidized ore to depths of 60 m. The oxidized ore is highly susceptible to leaching accompanied by migration down the dip of permeable sands. The unoxidized deposits form well-developed roll-type ore bodies along oxidation-reduction boundaries.

The main source of uranium appears to be the tuff of non-marine post-Eocene formations of the region. The tuff apparently came from the volcanic regions of New Mexico and West Texas and was blown or washed, together with erosion debris, into south Texas. The uranium, with some other minerals, was leached from the volcanic ash by oxygenated and alkaline waters, within an arid climate, through diagenesis and weathering. Carried downward through streams and channels as well as other permeable rocks, the uranium arrived at deducing environments below the water table. Possible H_2S and other gases precipitated the uranium by reduction, probably along fault lines. Deposits formed along the edges of strata of permeable host rocks interbedded with claystones and siltstones of low permeability Faults parallel to strike served to localize the deposits.

During the deposition of the uranium host rocks, the conditions were similar to those existing along the present Gulf Coast of Texas, except that the climate was less arid and there was a greater supply of volcanic debris (See, e. g., Dickinson, 1974).

Other details of this deposit are in the literature (See, e. g. Dickinson, 1974; Eargle et al., 1971; Eargle and Weeks, 1973; and Eargle et al., 1975). See also Clay West.

Tonco-Amblayo (Argentina).

This deposit, in the Province of Salta, contains uranium concentrations of 0.2% U_3O_8 (See, e. g., Stipanicic, 1970). The deposits are in continental sandstones, within a vestige of seismotectonomagmatic-belt activity that extends toward Peru and Bolivia (Tatsch, 1973a).

Tono (Japan).

See Tsukiyoshi.

Tsukiyoshi (Japan).

The Tsukiyoshi deposit is within the Tono mine area of the Gifu Prefecture. Other deposits in that area include Jorinji, Misano, and Utozaka. These deposits occur in the basal part of the Miocene Toki group. These are distributed in the tributaries or at the head of channels on the plane of unconformity under the formation. The typical ore mineral is a zeolite of the heulandite-clinoptilolite group with uranium adsorbed thereto.

The conduit of uranium-bearing ground waters that migrated from the basement granites into the Tertiary sediments was controlled by impermeable barriers of montmorillonite, by Tsukiyoshi fault, and by channel structures (See, e. g., Katayama et al., 1974). Uranium was adsorbed in zeolite when the uraniferous ground waters became stagnant.

Further uranium enrichment was produced by the oxidation of pyrite to produce sulfuric acid, which, in turn, first leached the uranium from the zeolite and then permitted it to be readsorbed by the zeolite. Resulting concentration of uranium reached 0.9%. Coffinite has formed where uranium was accumulated beyond the adsorption cap of the zeolite, or where strongly reducing conditions were maintained be carbonaceous matter.

Other details of this deposit are contained in the literature (See, e. g., Katayama et al., 1974; Hayashi, 1965, 1970).

<u>Tunisia</u> (Africa).

Like Morocco (q. v.).

<u>Urgeirica</u> (Portugal).

See Serra da Estrada.

<u>U. S. S. R.</u> <u>Sandstones</u> (Eurasia).

Many Tertiary vestiges of seismotectonomagmatic-belt activity in the U. S. S. R. contain economic deposits of uranium (See, e. g.,

Danchev et al., 1969; Tatsch, 1973a). The host rocks are mainly gravelstones, sandstones, and siltstones, reminiscent of similar Tertiary uraniferous deposits in other parts of the world. The host rocks fill intermontane basinal seismotectonomagmatic-belt vestiges surrounded by Hercynian mountains with granite and granodiorite intrusions in metamorphic rocks. Associated rhyolite flows are younger than is the granite. Both the granite and the rhyolite appear to have contributed uranium; they contain up to 0.005% U_3O_8. The sediments are dark and rich in coalified plant tissues. Very small amounts of copper, molybdenum, and vanadium accompany the uranium. There are traces of beryllium, chromium, gallium, germanium, lead, and zinc; but antimony, arsenic, bismuch, mercury, and tungsten are almost completely absent. The uranium is within sediments that accumulated during later stages. Diagenesis appears to have controlled the fixation of the uranium. The ultimate source of the uranium was the dikes associated with Mesozoic-Hercynian episodes of seismotectonomagmatic-belt activity (Tatsch, 1973a).

Other details of this are in the literature (See, e. g., Modnikov and Lebedev-Zinov'yev, 1969; Danchev et al., 1969).

Utozaka (Japan).

See Tsukiyoshi.

Verna Lake (Saskatchewan)

See Middle Lake.

Walvis Bay (Africa).

Four separate basins on the continental shelf off the coast of South West Africa conainn uraniferous muds with concentrations up to 100 ppm (= 0.01%) uranium. The best developed of these is near the Walvis Bay - Swakop River area. Similar smaller bays exist at the Unjob - Hoanib Rivers (20°S), north of Hallams Bird Island (24°S), and opposite Naribus (25^oS).

The uraniferous mud is generally olivine-green-gray, soft, and homogeneous, with a smell of H_2S. The mud is extremely fine-grained and comprises roughly 65% siliceous matter plus miscellaneous foraminifera, ostropods, pteropods, gastropods, lamellibranches, bentonite, and sharks' teeth. Similar uraniferous muds occur in other basins, including those of the Baltic, Black, and Caspian seas. The African muds average about 20 ppm uranium, roughly twice that of the Eurasian basins.

Webb County (Texas).

Similar to Felder (q. v.).

White River (Wyoming).

See Gas Hills and Shirley Basin.

Wind River (Wyoming).

See Gas Hills and Shirley Basin.

Wyoming (U. S. A.).

The intermontane basins of Wyoming contain uraniferous deposits derived from vestiges of seismotectonomagmatic-belt activity (Tatsch, 1973a). These vestiges are largely (1) granitic terranes and (2) volcanic tuffs.

In some cases, the same episodes of seismotectonomagmatic-belt activity that produced the unaniferous marine black shales of the Paleozoic also produced the vestiges from which the Tertiary sediments were evolved after a "reworking" of the older vestiges.

Other details of this deposit are contained in the literature (See, e. g., De Nault, 1975; Dodge and Powell, 1975). See also Shirley Basin.

Yeelirrie (Western Australia).

This area is a semi-arid part of a vestige of seismotectonomagmatic-belt activity in the form of a plateau underlain by Archean granite and associated rocks (Tatsch, 1973a). During the Tertiary, this vestige was deeply eroded by a major drainage system to the southeast into the Great Australian Bight. The mineralization occupies sediments of calcrete (caliche), clay, and sand. The uranium was deposited, as carnotite, from groundwater percolating through a filled ancient river bed. It occurs as thin films in horizontal layers lining the voids and cavities along 45 km of the channel. These are rarely more than 8 m below the surface. The average grade is about 0.15% U_3O_8.

Other details are included in the literature (See, e. g., Langford, 1974).

Zletovska Reka (Yugoslavia)

This deposit, located in the Osgovo Mountain area of Macedonia, corresponds genetically to hydrothermal mineralization formed in an epithermal environment. It appears to have originated from a deep magmatic source related to the final phase of the metallogenic epoch of the Alpine seismotectonomagmatic-belt episode (Tatsch, 1973a). The rocks comprise andesites, dacites, and related pyroclastics associated with the most extensive seismotectonomagmatic-belt volcanics of Yugoslavia.

Part of the Zletovo-Kratovo area, this deposit is structurally part of the large Dinarides system. The geology represents vestiges of several episodes of complex seismotectonomagmatic-belt activity. The basement comprises Paleozoic crystalline schists, some mica-schists and phyllites, and subordinate amphibloites and gneisses. In the area of Zletovska Reka, the basement is covered by extensive rocks.

The main structure of the area comprises fault-type structures, produced because the Zletovo-Kratovo area was inside a highly-mobile zone of the Alpine seismotectonomagmatic belt (Tatsch, 1972a). During this phase of structural development, the fold-type structures were replaced by higher-energy, fault-type structures and with associated magmatic activity expected within an active seismotectonomagmatic belt.

The most important faults are roughly northnorthwest-southsoutheast and orthogonal thereto (i. e., roughly eastnortheast-westsouthwest), corresponding to the directions of the two intersecting seismotectonomagmatic belts in that area at that time (Tatsch, 1972a). The first of these (that is, northnorthwest-southsoutheast) appears to have served as feeders for hydrothermal solutions with sulfide mineralization. Lead and zinc sulfide mineralizations are located in these structures. The second of these (that is, eastnortheast-westsouthwest) were produced by subsequent seismotectonomagmatic-belt activity. The uranium mineralization is located primarily within the eastnortheast arm of this fault system.

The ore body comprises uranium minerals, associated sulfides, and gangue minerals. Pitchblende is the most important uranium mineral. It forms stains, veinlets, and fine disseminations. The pitchblende is younger than the sphalerite and pyrite. Oxidation has converted some of the pitchblende to the secondary uranium minerals, including autunite ($Ca(UO_2)_2(PO_4)_2 \cdot 10\text{-}12\ H_2O$), kasolite ($Pb(UO_2)SiO_4 \cdot 8\text{-}12\ H_2O$), uranophane ($Ca(UO_2)_2SiO_7 \cdot 6\ H_2O$), torbernite ($Cu(UO_2)_2(PO_4)_2 \cdot 6\ H_2O$), and other phosphates and silicates. The zone of secondary mineralization lenses out at depth.

The associated sulfides include pyrite and sphalerite, plus some bravoite, galena, marcasite, and tetrahedrite. The gangue minerals include barite, chalcedony, fluorite, and quartz.

The Zletovska Reka uranium deposit was formed during a separate (and youngest) phase of the mineralizing processes associated with the Tertiary seismotectonomagmatic-belt activity in the Zletovo-

Kratovo area (Tatsch, 1973a). The deposition was influenced by vestiges of earlier seismotectonomagmatic-belt episodes that had produced intensive fault-type structures within the volcanic complex. Some of these structures, if suitably located, were later mineralized when the volcanites were consolidated and faulted, thereby providing paths of preferential flow for the circulation of hydrothermal solutions near the surface during the magmatic phase of the seismotectonomagmatic episode (Tatsch, 1972a).

During the post-magmatic activity, the copper, lead, uranium, and zinc deposits formed. It appears that all the mineralizing soltuions originated from the same, deep-seated magmatic source, formed during the final phase of the Alpine metallogenetic epoch (See, e. g., Radusinovic, 1974). The successive development of the magmatic and tectonic phase of the seismotectonomagmatic-belt activity caused characteristic zoning within the deposit, whereas the uranium deposition occurred after the other metals had been deposited. Consequently, the uranium was concentrated mainly in the younger westnorthwest-eastsoutheast fault structures, whereas the other metals were concentrated mainly in the earlier northwest-southeast structures. Laterally, the copper mineralizations are in the western part of the arm, those of the lead and zinc in the central part, and those of uranium in the eastern part. Some traces of gold appear west of the copper mineralizations.

Conclusions Regarding the Cenozoic Uranium Deposits.

Close scrutiny of the Cenozoic uranium deposits reveals that almost all of them appear to be associated with plate boundaries, rift valleys, extrusive and intrusive activity, metamorphism, hydrothermal activity, and other surficial manifestations of seismotectonomagmatic-belt activity that occurred within the upper 1000 km of the Earth during the Cenozoic. This may best be visualized by comparing the loacations of the actual deposits with surficial manifestations of the seismotectonomagmatic belts that have been active during the Cenozoic (See, e. g., Figure 2-8, page 34). In doing this, it is well to recall that the dashed lines shown on the figure are roughly 1000 km beneath the surface, whereas the uranium deposits are at or near the surface. Because the projection of energy through 1000 km of heterogeneous material over a period of 4.6 b. y. is not a straight-line function (See Chapter 2), the projections must be made realistically in terms of the Earth's actual geometrical, mechanical, thermal, and chemical heterogeneties (See, e. g., Tatsch, 1972a).

Chapter 7

MESOZOIC SEISMOTECTONOMAGMATIC BELTS AND THE ASSOCIATED URANIUM DEPOSITS

The uranium deposits associated with the Mesozoic seismotectonomagmatic belts comprise, according to the Tectonospheric Earth Model concept, part of the Earth's "original" early-Archean uranium that has been "reworked" by subsequent seismotectonomagmatic-belt activity during the Archean, the Proterozoic, and the Phanerozoic through the Mesozoic but not including any such activity during the Cenozoic. These latter deposits have been considered in Chapter 6.

Representative Mesozoic Uranium Deposits.

In order to determine how well the Earth's uranium deposits follow the patterns predicted by the Tectonospheric Earth Model, it is well to consider the locations and salient characteristics of representative examples of the Earth's known Mesozoic uranium deposits. It is not too surprising that these deposits are found primarily in those areas of the Earth's surface where vestiges of Mesozoic seismotectonomagmatic-belt activity still remain today (Tatsch, 1973a).

Ambrosia Lake (New Mexico).

Most of the ore in this district is strata-bound, peneconcordant, layered, humic impregnations containing extensive accumulations of uranium (See, e. g., Granger et al., 1961). However, some roll-type deposits occur in some parts of the district where one or more altered tongues have encroached on the earlier tabular deposits to produce a redistribution of the ore bodies.

The clay mineralogy of this deposit appears to have been influenced by the uranium mineralization and oxidation of the Morrison Formation (Jurassic) of the area (See, e. g., Lee et al., 1975).

This suggests that a study of the clay-sized fractions should set constraints on the origin and evolution of sandstone-type deposits.

Within the district, the altered hematitic core ranges from pale to intense red. In most places, the core is surrounded by a buff-to-orange limonitic rock as much as 100 m thick near the roll front. Near the contact with reduced rock, some pyrite-derived goethite pseudomorphs are found. Other details are contained in the literature (See, e. g., Granger and Warren, 1974; Granger et al., 1961; Lee et al., 1975). This deposit is part of the Grants Mineral Belt (q. v.).

Andujar (Spain).

These uraniferous sandstones average better than 0.10% U_3O_8.

Araxa (Brazil).

These deposits are similar to Conway (q. v.) but they are more concentrated and less extensive and occur within pyrochlore bearing alkalic rocks.

Arino-Alloza-Alcorisa (Spain).

These deposits lie in very complex terrain. The uranium content is roughly 0.01% U_3O_8. They are similar to Picofrentes (q. v.).

Austen (Nevada).

This area comprises mid-Mesozoic quartz monzonite in contact with metamorphosed sedimentary rocks with minor silicified limestone and hornfels in the Reese River silver district about 5 km south of Austen. Austinite and metatorbernite occur within faults and fractures of both the intrusive and the sediments. Some uranium is disseminated in small roof pendants. Narrow siliceous dikes cut the monzonite. Numerous mineralized fracture zones suggest ascending uranium mineralizing solutions. The uranium minerals are associated with copper, gold, iron, and silver in gnague of sericite and some vein quartz. Selected samples contain as much as 6% U_3O_8; grab samples average about 0.15%.

Austrian Alps (Europe).

Some of the Permo-Triassic sandstone uranium deposits of the Austrian Alps were mobilized by Alpine seismotectonomagmatic-belt activity (Tatsch, 1973a). See also Chapter 8.

Baja California (Mexico).

The states of Baja California and Baja California Sur contain many vestiges of seismotectonomagmatic-belt activity that appear to be uraniferous (Tatsch, 1973a).

Barruelo de Santullan (Spain).

This deposit is similar to Mazarete (q. v.).

Big Indian (Utah).

This deposit is located in Lisbon Valley, northeast of the deposit at Monument Valley (q. v.), and southwest of the deposit at Uravan (q. v.). It is in the Moss Back member of the Chinle formation, i. e., within the lowest member of that area. It comprises sandstone and mudstone, plus a few fragments of coalified fossil flora. The Moss Back was deposited rapidly within a northwest-trending vestige of seismotectonomagmatic-belt activity (Tatsch, 1973a). The Big Indian belt is about 1 km wide and 25 km long parallel to that vestige, which was produced during the late late Paleozoic and in association with the Laramide episode of the late Mesozoic.

Other details are contained in the literature (See, e. g., Fischer, 1974b).

Bohemia (Europe)

See Czechoslovakia.

Bugaboo Creek (British Columbia).

Gravels in the alluvial plains of Bugaboo Creek, southeast of Golden, contain uranium minerals glacially derived from a Mesozoic quartz-monzonite.

Chihuahua (Mexico).

See Sierra Pena Blanca.

Coahuila (Mexico).

The State of Coahuila contains many vestiges of seismotectonomagmatic-belt activity that appear to be uraniferous (Tatsch, 1973a).

Colorado Plateau (U. S. A.).

These low-grade (0.05% U_3O_8) deposits occur within several sandstone horizons of the Morrison Formation. They appear to have been derived from granite vestiges of seismotectonomagmatic-belt activity (Tatsch, 1973a). The main uranium minerals are coffinite and pitchblende. Deposits under this heading include those at Grants (New Mexico), Monument Valley - White Canyon (Utah), Big Indian (Colorado), and Uravan (Colorado). These are discussed under those headings.

Members of the Morrison Formation on the Colorado Plateau were deposited in large subaerial fans in a broad northeasterly-trending trough-like vestige of seismotectonomagmatic-belt activity (Tatsch, 1973a). This vestige existed between the low Uncompahgre - San Juan highlands (southwestern Colorado) and the taller Mongollon highlands (southern Arizona). The Morrison detritus came from the Mongollon vestige. This deposition represents a shift of the drainage pattern to the northeast and east from the erstwhile west and northwest patterns (See, e. g., Saucier, 1975). The progressive tilting of the paleoslope to the northeast persisted through the Cretaceous.

The Morrison Formation comprises mainly reworked sedimentary rocks, deeply-weathered volcanic material, and some metamorphic constituents. The acid volcanics were derived from southern Arizona, where hundreds of m of rhyolitic and dacitic tuffs and flows of late Triassic and Jurassic ages have been preserved in what appears to be northwesterly-trending graben-like vestiges of seismotectonomagmatic-belt activity (Tatsch, 1973a).

The pronounced northwesterly-trending pre-Laramide tectonic grain of the Colorado Plateau and surrounding areas significantly affected the Morrison sediments (See, e. g., Saucier, 1975). This, in turn, controlled the present mineralization.

It appears that trends of the paths of preferential flow within the sandstone bodies save been determined by broad, shallow folding that was associated with the applicable seismotectonomagmatic-belt activity contemporaneous with the Morrison deposition. These trends of the preferred zones of transmission then controlled the movement of the ground waters and, ultimately, the location and trends of the uranium orebodies. Other details are included in the literature (See, e. g., Downs and Runnells, 1975; Saocier, 1975).

Conway (New Hampshire).

This uraniferous igneous body contains anomalous concentrations of uranium as an indigenous or genetic constituent. The Triassic-Jurassic biotite phase of the Conway Granite averages about 0.0015% U_3O_8, over an area of roughly 800 km^2. It also contains about 0.0064% ThO_2. This granite vestige of seismotectonomagmatic-belt activity extends to a depth of about 300 m and contains roughly 200 km^3 estimated to contain over 7 million tons of U_3O_8.

Cottian Alps (Italy).

See Preit Valley.

Cuenca (Spain).

See Mazarete.

Czechoslovakia (Europe).

The Mesozoic uranium deposits of Czechoslovakia are associated with vestiges of seismotectonomagmatic-belt activity. Some of these may be seen in the Cenomanian of Bohemia (See, e. g., Samana, 1973). These deposits occur in the terrigenous formations of the top of the late Cretaceous or the base of the upper Cretaceous.

The ores at Joachimsthal occur within fissure veins together with dolomite, quartz, silver sulfides, and some compounds of bismuth, cobalt, copper, iron, lead, nickel, silver, and zinc.

Fieberbrunn (Austria).

This Permo-Triassic uranium deposit is within gray sandstone in the northern graywacke zone of Tyrol. The sequence begins with coarse-grained standstone, followed by red ciricite quartz sandstone and sericite schist. Intercalated with these are carboniferous gray-green sandstone with hematite, pyrite, and marcasite. Some floral relicts are present within the gray sandstone. Some horizons contain apparently synsedimentary copper enrichments plus some pyrite and pitchblende. The uranium ores, together with pyrite, fill the cell walls and pores of the carbonized floral remains.

The Triassic mineralization occurred in lagoonal-brackish, shallow water. The uranium appears to have been precipitated syndiagenetically from metalliferous weathering solutions when they encountered reducing conditions during the sedimentation. These deposits are similar to the late Permian deposits of Italy and Yugoslavia (Chapter 8). Other details are in the literature (See, e. g., Barthel, 1974).

Forez (France).

These uraniferous sandstones, averaging better than 0.10% U_3O_8, are part of the Bois Noris - Limouzat vein of the Central Massif. The mineralized vein appears to have been developed in six stages:

a. The deposition of quartz-marcasite-pitchblende and pyrite.

b. The deposition of some hematitic quartz. At this time, the pitchblende was slightly altered into coffinite, and the epidote-adularia spread in the surrounding rocks, probably as the result of an episode of seismotectonomagmatic-belt activity (Tatsch, 1973a).

c. The banded smoky quartz, fluorite, and carbonates were formed.

d. The whewellite was formed.

e. A late generation of quartz-bismuthinite intruded the early generations.

f. Supergene alteration occurred during Alpine seismotectonomagmatic-belt activity and since then.

Close scrutiny of these six stages suggests a complex history, with wide variations in temperatures and compositions. Rich concentrations of CO_2 in the fluids associated with the pitchblende deposition and with the bismuthinite suggests that the uranium was transported through uranyl carbonate complexes (See, e. g., Poty et al., 1974).

France (Europe).

Some French uranium deposits are Jurassic (158 m. y.) but most are Permian (230 m. y.). See Chapter 8.

Golden (British Columbia).

See Bugaboo Creek.

Gorenja Vas (Czechoslovakia).

See Chapter 8.

Grants (New Mexico).

Grants represents one of the few cases in the U. S. A. where uranium ores are associated with limestone. The ore occurs in the Jurassic Todilto limestone. These deposits are concentrated wherever the limestone has been deformed locally by intraformational folds and intensely fractured by vestiges of Jurassic seismotectonomagmatic belt activity (Tatsch, 1973a). Underlying the Todilto is a thick permeable eolian sandstone (Entrada formation) comprising an upper zone of bleached sandstone about 10 m thick lying between the Todilto contact and the main Entrada red bed.

The presence of some fluorite suggests that the Todilto uranium ores were deposited from solutions having a magmatic hydrothermal component. The absence of faults, however, precludes access for the mineralizing solutions except through the overlying limestone or through the underlying sandstones. The transport mechanism was probably ground water, becasue the Summerville formation above the Todilto appears too impermeable to have been an aquifer.

The involvement of reducing conditions in the formation of the Grants uranium deposits are reminiscent of similar conditions that produced the Kupferschiefer mineral deposits of central Europe (See, e. g., Tatsch, 1975a). Other details of the Grants deposits are covered in the literature (See, e. g., Adler, 1974; Moench and Schlee, 1967; Anderson and Kirkland, 1960; Brookings and Lee, 1974;

Brookins, 1975; Lee et al., 1975; Rawson, 1975; Lapp, 1975).

Grants-Laguna (New Mexico).

See Grants.

Green River (Utah).

The Morrison formation of the Green River district of southeastern Utah is being tested to determine the origin, evolution, and present characteristics of the uranium contained therein (See, e. g., Anon., 1975b). Initial analyses suggest that the uranium in the buried Morrison is traceable to the same seismotectonomagmatic-belt activity that is responsible for other uranium deposits of that time (Tatsch, 1973a).

Hamr (Czechoslovakia).

See Czechoslovakia. This deposit lies within Cretaceous sediments of the Bohemian Massif.

Hecla (Utah).

This deposit is part of the Colorado Plateau (q. v.).

Hopi (Arizona).

See Navaho.

Italian Metamorphic Rocks (Italy).

During the metamorphism associated with the Alpine episode of seismotectonomagmatic activity, some of the Permian uraninite and non-ferrous metals of northern Italy were mobilized and underwent complex recrystallization (See, e. g., Tatsch, 1973a). This caused some of the older ores to migrate to neighboring rocks. See also Chapter 8.

Jackpile (New Mexico).

This is part of the Grants Mineral Belt (q. v.).

Joachimsthal (Czechoslovakia).

See Czechoslovakia.

Julianehaab (Greenland).

This is similar to Conway (q. v.) but more concentrated and less extensive. It occurs in solalite foyaite and nepheline syenite.

Khorat Plateau (Thailand).

See Phu Wieng.

Konigstein (Germany).

This East Germany deposit lies within the Bohemian Massif sediments. See Czechoslovakia.

Limousin (France).

These uraniferous sandstones, averaging better than 0.10% U_3O_8, are part of a vein of the Central Massif. The uranium comprises, basically, two forms: (1) veins and (2) hydrothermally altered pods of the Saint-Silvester two-mica granite. Close scrutiny of the second type suggests a clear relation between uranium grade and CO_2, that boiling occurred at the time of the trapping, and that deposition occurred at about $350^{o}C$ and roughly 750 bars (See, e. g., Poty et al., 1974). These observations suggest that the mineral was transported as uranyl carbonate complexes and deposited due to unmixing of the solution. Reduction of the uranyl ion was probably controlled by a constituent of the solution. A reaction involving the oxidation of ferrous iron does not seem adequate because hematitic alteration is not widespread.

Lisbon (Utah).

Located near Moab, in the Big Indian district, this Rio Algom deposit produces more than 500 tons U_3O_8 per year (See, e. g., Jackson, 1975). This is the first uranium mine northeast of the Lisbon fault, suggesting that this area may resemble the deposits on the southwestern side (e. g., Hecla, Mi Vida, North Alice, and Radon).

The Lisbon Valley anticline, part of the Colorado Plateau (q. v.), was arched by the upwelling of salt and gypsum in the Hermosa formation during late Paleozoic and Larimide episodes of seismotectonomagmatic-belt activity (Tatsch, 1973a).

Mazarete (Spain).

These deposits are in Triassic sands of the Bunandstein type. In many respects, they resemble those of Utah (q. v.). The formation is lower Triassic red sands, part of which has been bleached. There is some dispersed organic material within the bleached sections. Other deposits that resemble this include Barruelo de Santullan, Reinosa (Santander), Valdemeca (Cuenca), and Agreda (Soria).

Mendoza (Argentina).

These uraniferous sandstones contain roughly 5000 tons averaging better than 0.10% U_3O_8.

Mequinenza (Spain).

These deposits are similar to those in Palencia (q. v.).

Mexico (North America).

See Baja California, Chihuahua, Coahuila, Nuevo Leon, Oaxaca, Sonora, and Tamaulipas.

Midnite (Washington).

These uraniferous epigenetic deposits probably formed soon after Cretaceous granitic intrusions into Precambrian metasedimentary rocks. Individual ore bodies are 60 m wide, 200 m long, and 50 m thick. They probably formed from hydrothermal solutions directed along favorable zones adjacent to the intrusive contact. The average mined grade is 0.23% U_3O_8.

Several open-pit mines have been developed in this area about 80 km northwest of Spokane. The uranium occurs at the contact between regionally metamorphosed Precambrian sediments and a granite that was intruded during an episode of Cretaceous seismotectonomagmatic-belt activity (Tatsch, 1973a). The intrusive is a coarse-grained, holocrystalline rock with aplitic and pegmatitic phases but without any discernible uranium mineralization. The uranium appears to be only at the contact and within and near associated fractures and shear zones. Secondary oxidized minerals (e. g., autunite and meta-autunite) lie above a fluctuating water table. A zone of partially oxidized, sooty, and compact uraninite with associated sulfides is below the water table. Hydrothermal action is indicated by the presence of adularia, illite, kaolinite, and montmorillonite. The average uranium content is about 0.285% U_3O_8 (Von Backström, 1974b).

Mi Vida (Utah).

This is part of the Colorado Plateau (q. v.).

Monument Valley - White Canyon (Arizona and Utah).

These Colorado Plateau deposits are in the Late Triassic Chinle formation of northern Arizona and southern Utah (See, e. g., Fischer, 1974b). They resemble the Big Indian (q. v.) deposits to the northeast. The host rock is mainly brightly-colored mudstone and siltstone, plus some clay and sand. The clay appears to have been derived from volcanic debris. Fragments of coalified fossil flora are present in most of the sandstone beds, especially the conglomeratic ones. Most of the ore deposits are in the lower parts of these conglomeratic sandstones. Most bodies are tabular elongated parallel to the host channel and related vestives of seismotectonomagmatic-belt activity (Tatsch, 1973a). The mineral belt is 2 to 4 km wide and just over 200 km long. The main minerals are copper, uranium, and vanadium. These appear to have been leached from the Shinarump member of the Chinle. Transport was by surficial and ground waters along the Shinarump vestiges of seismotectonomagmatic-belt activity to fossil-wood areas where precipitation occurred. The miniralization appears to be late Triassic or early Jurassic; but some evidence suggests Cretaceous (See, e. g., Fischer, 1974b).

Morvan (France).

These uraniferous sandstones average better than 0.10% U_3O_8.

Mount Taylor (New Mexico).

This deposit, part of the Grants Mineral Belt (q. v.), lies midway between Ambrosia Lake (q. v.) and Jackpile (q. v.). Drilling suggests that this deposit contains at least 100 million lb of U_3O_8 (See, e. g., Lapp, 1975).

Nacimiento-Jemez (New Mexico).

The uranium in this north-central area of New Mexico (i. e., Sandoval and Arriba counties) is present in rocks that span the Pennsylvanian-to-Tertiary age. Deposits conprise uranium minerals disseminated in sandstone, siltstone, carbonaceous shale, coal, and limestone in association with carbonized plant debris in sandstone and red-bed copper deposits. The minerals fill the interstices of a silicified rhyolite breccia in one area. The source of the uranium appears to be the overlying volcanic rocks of the Pleistocene Bandelier Tuff.

New Jersey (U. S. A.).

See Pennsylvania.

Nigeria (Africa).

This deposit is similar to Conway (q. v.) but it is more concentrated and less extensive. It occurs in riebeckite granite.

Nisa (Portugal).

These uraniferous sandstones average better than 0.10% U_3O_8.

North Alice (Utah).

This is part of the Colorado Plateau (q. v.).

Oaxaca (Mexico).

The State of Oaxaca contains many vestiges of seismotectono-magmatic-belt activity that appear to be uraniferous (Tatsch, 1973a).

Ontario (Canada).

The deposits of Ontario are similar to Conway (q. v.), but they are nore concentrated and less extensive. The occur in pyrochlore-bearing alkalic rocks.

Parana Basin (Brazil).

This area of southern Brazil contains both Mesozoic and Paleozoic deposits. These are discussed in Chapter 8.

Penalen (Spain).

This deposit, situated in Guadalajara, contains about 0.01% U_3O_8 in sand. It is similar to Picofrentes (q. v.).

Pennsylvania (U. S. A.).

The uraniferous Triassic basins of eastern Pennsylvania and

northern New Jersey are being tested to determine the source of their uranium (See, e. g., Anon., 1975b). Initial analyses suggest that this uranium is traceable to the Applachian vestiges of seismotectonomagmatic-belt activity in that area (Tatsch, 1973a).

Phosphoria (U. S. A.).

See Chapter 8.

Phu Wieng (Thailand).

The Phu Wieng sandstone uranium-copper deposit of the Khorat Plateau is similar to the uraniferous deposits of the Colorado Plateau (q. v.). The host rock appears to be carbonaceous fluvial sandstones and conglomeratic sandstones of the Jurassic vestiges of the Triassic -to-Cretaceous episode of seismotectonomagmatic-belt activity (Tatsch, 1973a). This part of the northeastern area of Thailand is structurally and stratigraphically similar to the Colorado Plateau. The uranium-copper mineralization appears to be closely associated with the red beds of the Khorat Plateau (See, e. g., Shawe et al., 1975). The uranium content is about 0.03% to 0.08% U_3O_8.

Picofrentes - Sierra Llana (Spain).

These deposits of the Iberian Cordillera are near Soria in the northeastern part of Spain. The mineralization, containing roughly 0.01% U_3O_8, is in the lower Cretaceous. The mineralizing agent was circulating solutions aided by supergene enrichment. Other deposits similar to this include Arino-Alloza-Alcorisa, Penalen, Salas de Infantes, and others in those areas. In many respects, these deposits resemble those of Wyoming (q. v.).

Placerville (Colorado).

See Rifle Creek.

Portugal (Europe).

Portugal's uranium deposits are associated with seismotectonomagmatic-belt activity dating back to the Precambrian. They appear to be supergene and related to the contact between Precambrian-Paleozoic metasediments and Hercynian granites. The uranium always occurs within the contact metamorphism aureole and rarely deeper than 50 m. Mineralization occurs in fractures near the granite roof and walls in metasediments. The brecciated structures appear to be vestiges of seismotectonomagmatic-belt activity that originated under tangential tectonic movements (See, e. g., Tatsch, 1972a). This caused shear faults and mylonitization without recrystallization of the rock elements. The most common uranium minerals are coffinite and pitchblende. Secondary minerals include autunite, sabugalite, torbernite, uranocirate, and small amounts of uranophane, parsonite, phoshuranylite, uranopilite, and zippeite. See also Spain.

Preit Valley (Italy).

During the metamorphism associated with the Alpine seismotectonomagmatic-belt activity, some of the Permian uraninite and nonferrous metals of the Preit Valley were mobilized and underwent complex recrystallization (See, e. g., Tatsch, 1973a). This caused some of the older ores to migrate to neighboring rocks. See also Chapter 8.

Pryor Mountains (Montana-Wyoming).

These deposits are similar to those at Grants (q. v.), but they are poorer in unanium content.

Radon (Utah).

This is part of the Colorado Plateau (q. v.).

Rifle Creek (Colorado).

This deposit, in Garfield County, lies within the Jurassic Entrada formation, a highly sandy eolian red bed extending over a large area of the Colorado Plateau. Besides the Rifle Creek area, uranium deposits occur also within the Placerville area of San Miguel County, Colorado, and in the Grants-Laguna area of northwestern New Mexico. Found in association with vanadium, the uranium of this area is confined to the bleached parts of the normally red rocks.

The bleaching of the Rifle Creek area includes the entire thickness of the Entrada sandstone plus the underlying Navajo sandstone and into part of the Triassic Chinle formation. In the Placerville and Grants areas, the bleaching is restricted to the upper few meters of the Entrada formation. The cause of the bleaching is not completely understood (See, e. g., Adler, 1974).

The bleached zone comprises white, limonitic, and some greenish layers plus some irregularly shaped zones. Other details have been discussed in the literature (See, e. g., Moench and Schlee, 1967). There is some evidence that the bleaching predates the Jurassic folding and may be related to the deposition of the overlying Todilto limestone. The obliqueness of the bleached boundary to the bedding suggests a geochemical rather than sedimentological origin for the bleaching.

The Rifle Creek ore in the Entrada and Navajo formations comprises three limbs forming a large flattened S-shaped body. Interruptions within the limbs form a series of smaller ore bodies showing internal roll-like features. A persistent characteristic of the Rifle Creek and Placerville limbs is the zoning of the metals, including V, U, Pb, and Cr. Fischer (1960) proposed that raaction at the interface of two solutions might account for element fractionation as well as for formation of the ore layers.

The deposits at Rifle Creek suggest a far more extensive

reducing zone than that which would have been inherent in the simple diffusion of H_2S from a Morrison-age sea floor into the subjacent water-saturated red beds. Fischer's (1960) analysis indicates, however, that the bleaching surrounding the ore was produced by H_2S reduction associated with petroleum that had extensively penetrated the pre-Morrison formation. Other details regarding speculations about the bleaching of the ore zone may be found in the literature (See, e. g., Adler, 1974).

Ross-Adams (Alaska).

The Ross-Adams uranium-thorium deposit is located within a small stock of alkalic granite of late Cretaceous or early Tertiary age. The country rocks are Devonian sediments and volcanics, black slate and hornfels, diorite, granodiorite, aplite, and alaskite. The youngest intrusive rock, excluding dikes, is alakli granite. This forms an 8 km^2 stock containing unusual quantities of certain minor elements, particularly cerium, lanthanum, niobium, thorium, uranium, yttrium, and the rare earths.

Both in shape and mineralogy, the Ross-Adams orebody represents an uncommon type of uranium deposit. No less unusual is its affiliation with alkali granite. The main ore minerals are uranothorite and uranoan-thorianite. The core of the orebody contains more than 0.5% U_3O_8. This is envoloped by a lower-grade uraniferous zone (Von Backström, 1974b).

Russia (Eurasia).

See U. S. S. R.

Sa-byr-Say (U. S. S. R.).

These uraniferous sandstone deposits are reminiscent of those at Mecseh (q. v., in Chapter 8).

Saint Bernhard (Switzerland).

During the metamorphism associated with the Alpine seismo-tectonomagmatic-belt activity, some of the Permian uranium and non-ferrous metals of this area were mobilized and underwent complex recrystallization (See, e. g., Tatsch, 1973a). This caused some of the older ores to migrate to the neighboring rocks. See also Chapter 8.

Salamanca (Spain).

See Spain.

Salas de los Infantes (Spain).

These deposits are similar to Picofrentes (q. v.), and they probably embody an extension of the same vestiges of seismotectono-magmatic-belt activity. Copper oxides are also present.

Salta (Argentina).

These uraniferous sandstones contain roughly 5000 tons of minerals averaging better than 0.10% U_3O_8.

Sandoval County (New Mexico).

Northeast of the Laguna district, in New Mexico, both primary and secondary uranium mineralization is present in the Westwater Canyon member of the Morrison formation. The Westwater channel sands appear to have originated in vestiges of seismotectonomagmatic-belt activity in that area (Tatsch, 1973a). During (and possbily after) the deposition of these sands, uraniferous solutions were instrumental in precipitating coffinite (See, e. g., Hafenfeld and Brookins, 1975). Later oxidizing ground waters appear to have destroyed the pyrite and calcite that were associated with that vestige. Some of the uranium was redeposited as secondary or "stock" ore, leaving a stain of iron-oxide on the host rock. Cementation was by kaolinization.

San Juan (New Mexico).

See South San Juan.

Santa Coloma de Queralt (Spain).

These deposits are similar to those in Palencia (q. v.).

Santander (Spain).

See Mazarete.

Sierra Llana (Spain).

See Picofrentes.

Sierra Pena Blanca (Mexico).

This deposit, in the State of Chihuahua, is associated with uraniferous vestiges of seismotectonomagmatic-belt activity (Tatsch, 1973a). Estimated size is 3500 tons.

Slick Rock (Colorado).

These uranium-vanadium deposits are in continental red beds of Permian and Mesozoic age. They are underlain by reduced marine Pennsylvanian rocks, and overlain by Cretaceous rocks. The red beds are mostly oxidized, but two facies are reduced: (1) one related to abundant coalified plant material; and (2) the other related to epigenetic alteration. The ore deposits occur only in altered rocks in the presence of carbonaceous material.

Reducing solutions related to vestiges of seismotectonomagmatic-belt activity appear to have altered the red beds and deposited the ores (Tatsch, 1973a). The upper Cretaceous Mancos Shale appears to have supplied the reductant (See, e. g., Shawe, 1974). In many respects this deposit resembles Grants (q. v.).

Small Fry (Utah).

This deposit, near Moab, has reserves of about 400 tons of U_3O_8.

Sonora (Mexico).

The State of Sonora contains many vestiges of seismotectonomagmatic-belt activity that appear to be uraniferous (Tatsch, 1973a).

Soria (Spain).

See Picofrentes; Mazarete.

South San Juan (New Mexico).

The deposits at South San Juan basin comprise two distinct generations of uranium within the upper fluvial members of the Jurassic Morrison formation. These two generations appear to have been associated with the same vestiges of seismotectonomagmatic-belt activity (Tatsch, 1973a).

The first generation ore comprises mainly coffinite in a matrix of interstitial matter (See, e. g., Squyres, 1974). The ore bodies are elongated, tabular, or lenticular masses trending parallel to the stream channel and related vestiges of seismotectonomagmatic-belt activity. Agewise, the ore bodies are almost as old as the host rocks. They appear to have been deposited from ground water at shallow depth.

The second generation ore comprises a much younger derivative that was formed by redistribution of some of the earlier-formed deposits. These younger deposits are roughly of a roll-type and occur near the periphery of a zone of recent oxidation extending downdip from the outcrop.

Other details of this deposit are contained in the literature (See, e. g., Hafenfeld and Brookins, 1975; Sandusky, 1975).

Spain (Europe).

The uraniferous deposits of Spain are similar to those of Portugal (q. v.). At Salamanca, the Hercynian adamellite and granodiorite are associated with intrusive vestiges of seismotectonomagmatic-belt activity (Tatsch, 1973a). These extend into the Cambrian, and they are responsible for a thermal aureole together with spotted slates, hornfels, quartzite, and limestone found in that area. The mineralization occurs, along bedding and joint planes together with some organic material, in stockwork form. The origin appears to have been supergene and confined to the metamorphic aureole ranging from 200 to 1000 km wide. The uranium minerals are autunite, pitchblende, torbernite, and uranophane. As in the Portuguese deposits, there are also some secondary minerals, including ianthenite, phosphuranylite, renardite, saleeite, and uranopilite. There is also some chalcopyrite, galena, pyrite, and melnikovite. Gangue minerals are barite, calcite, jasper, and quartz.

Texas Caliche (U. S. A.).

The uraniferous caliche (calcrete) of west Texas appears to have formed within vestiges of late Mesozoic and early Cenozoic episodes of seismotectonomagmatic-belt activity. These are discussed under this heading in Chapter 6.

Texas Mesozoic (U. S. A.).

Most of the well-known Texas uranium deposits are in the Texas Coastal Plain (q. v.). But many less-well-known deposits occur, in other parts of Texas, within older vestiges of seismotectonomagmatic belt activity, dating from Pennsylvanian time through all of the Mesozoic and into the Cenozoic. Specific formations include the Tecovas and Trujillo (Triassic) and Edwards limestone (Cretaceous).

Thailand (Asia).

See Phu Wieng.

Tucano (Brazil).

This basin, in the State of Bahia, contains anomalous uranium and vanadium deposits associated with organic matter. In the lower part of the Cretaceous Sergi formation, these low-grade concentrations are associated with subsurface black minerals and surficial vanadates (See, e. g., De Andrade-Ramos and Fraenkel, 1974).

Turkey (Europe).

Some uranium veins and peneconcordant sandstones of Turkey average better than 0.10% U_3O_8.

Uchkuduk (U. S. S. R.).

These uraniferous sandstone deposits are reminiscent of those at Mecsek (q. v. in Chapter 8).

Uravan (Colorado).

The Uravan deposit is in the Salt Wash member of the Morrison formation of southwestern Colorado. The Salt Wash was deposited as a broad alluvial fan within a gently-dipping vestige of seismotectono-magmatic-belt activity that then covered the eastern third of Utah, the western third of Colorado, and the adjoining areas of Arizona and New Mexico (Tatsch, 1973a; Fishcer, 1974b). The host rock is mostly conglomeratic sandstone plus some mudstone. The uranium-vanadium mineralizations are mainly in the stream-laid sandstone facies. There are some coalified fossil flora. The deposits are tabular within a belt of roughly 110 km long and a few km wide.

Other details of this deposit are contained in the literature (See, e. g., Fischer, 1974b0.

Uttar Pradesh (India).

These Jurassic phosphorite nodules and pellets of fluorapatite and corbonate apatite contain 0.01% to 0.02% U_3O_8.

Valdemeca (Spain).

This deposit is similar to Mazarete (q. v.).

Vendee (France).

The mineralized body at Vendee comprises a vein. The wall rock is strongly hematitized. The vein is connected upward and disappears downward. The vertical development is relatively shallow. The mineralization, which is younger than is the country rock, appears to coincide with a period of peneplanation.

The uranium was concentrated by: (1) a selective endogenous leaching in the fractured zones, by the action of waters re-heated by a late episode of seismotectonomagmatic-belt activity (Tatsch, 1973a); (2) the descending infiltration of uraniferous meteoric waters; and (3) a quasi-lateral flow of a combination of connate and meteoric waters to redistribute the uraniferous materials.

This deposit has some characteristics reminiscent of those of the Wyoming (q. v.) roll-type deposits (See, e. g., Poty et al., 1974). The mineralization averages better than 0.10% U_3O_8.

West Germany (Europe).

Some uraniferous veins and peneconcordant sandstones of West Germany average better than 0.10% U_3O_8. Many of these are associated with vestiges of Hercynian episodes of seismotectonomagmatic-belt activity (Tatsch, 1973a).

White Canyon (Utah).

See Monument Valley.

Wyoming (U. S. A.).

The uraniferous sandstone deposits of the Wyoming intermontane basins resemble those of the Colorado Plateau (q. v.).

Conclusions Regarding the Mesozoic Uranium Deposits.

Close scrutiny of the Mesozoic uranium deposits reveals that almost all of them appear to be associated with those areas of the Earth's surface where vestiges of Mesozoic seismotectonomagmatic-belt activity remain today.

Chapter 8

PALEOZOIC SEISMOTECTONOMAGMATIC BELTS AND THE ASSOCIATED URANIUM DEPOSITS

The uranium deposits associated with the Paleozoic sesimtoectonomagmatic-belts comprise, according to the Tectonospheric Earth Model concept, part of the Earth's "original", early-Archean uranium that has been reworked by subsequent seismotectonomagmatic-belt activity during the Archean, the Proterozoic, and the Paleozoic, but not during the Mesozoic or Cenozoic. The Mesozoic and Cenozoic deposits have been considered in earlier chapters.

Representztive Paleozoic Uranium Deposits.

In order to determine how well the actual uranium deposits follow the patterns predicted by the Tectonospheric Earth Model, it is well to consider the locations and characteristics of representative examples of known Paleozoic uranium deposits. This is done in the following sections.

Adams (Mexico).

These peneconcordant uraniferous deposits in sandstone are located in the State of Chihuahua. They comprise roughly 2500 tons averaging better than 0.10% U_3O_8.

Alabama (U. S. A.).

These uraniferous black shales resemble those of Tennessee (q. v.).

Almorah-Pithoragarh (India).

See Kihtwar.

Armorinopolis (Brazil).

This new deposit, about 300 km southwest of Brasilia, has indicated reserves of 500 tons of ore assaying 0.02% to 0.5% U_3O_8.

Arkansas (U. S. A.).

These uraniferous marine black shales resemble those of the State of Tennessee (q. v.).

Austrian Alps (Europe).

Most of the uranium deposits of the Austrian Alps resemble those of other parts of the Alps, particularly in that they are associated with vestiges of earlier seismotectonomagmatic-belt episodes (Tatsch, 1973a). This is seen in the Permo-Skythic beds of the Eastern Alps (Petrascheck et al., 1974). The content of the ore varies from 0.05% to 2.0%. These beds are of fluvial, lagoonal, and shallow marine origin. The northernmost zone is represented by the Bundsandstein and Werfener Schiefer. In the Southern Alps, the Permian Grödener Sandstein, with its well-known intercalation of quartz porphyry, overlies upper Carboniferous gray shales and sandstones.

Some of the best deposits are at Forstau (Salzburg), at Mayrhofen (Tyrol), at Fieberbrunn (q. v.) (Tyrol), and at Mühlbach (q. v.) (Bischofshofen).

Other features of these deposits are described in the literature (See, e. g., Mittelberger, 1970; Grutt, 1971; Petrascheck et al., 1974; Siegl, 1972).

Azelick (Niger).

These peneconcordant uraniferous ores in sandstone contain over 5000 tons averaging better than 0.10% U_3O_8.

Bagadwani (India).

See Bartalao.

Barakar (India).

These lacustrine sediments contain calcified wood remains, clays, and coal seams. They are overlain by the uraniferous Motur (q. v.) and Bijori (q. v.) beds, but contain little or no uranium concentrations within their sandstones. Other details are contained in the literature (See, e. g., Udas and Mahadevan, 1974).

Bearpaw Mountain (Montana).

These uraniferous ores with syenite and associated pegmatite comprise relatively small bodies averaging about 0.05% U_3O_8 (roughly 500 grams per ton). The total tonnage is about 15,000 tons.

Big Meadow (Idaho).

See Idaho.

Bijori (India).

These fluvial sediments contain significant quantitits of uranium and thorium. The beds comprise pale green and buff-colored gritty and arkosic sandstones with pods and nodules of clay. They resemble the Moturs (q. v.) but are more carbonaceous and carry some fossil wood. Their silicified and carbonaceous aspects reflect the oxidation conditions during the diagenesis and consolidation of the sediments (See, e. g., Udas and Mahadevan, 1974). The minerals are mainly monzonite and zircon with ferruginous gritty sandstones, pebbly beds, and conglomerates. The U:Th ratio is high (as much as 1:40), reflecting the monzonite-zircon content.

Billingen (Sweden).

This deposit, at the Ranstad mine, is one of the world's largest. The ore grades only about 0.02% U_3O_8. The host rock is shale, which contains hydrocarbons. Byproducts include molybdenum, nickel, and vanadium. Planned production will be increased to 1500 tons/yr.

Bodipani (India).

Significant concentrations of uranium occur within this part of the Motur (q. v.) bed. The uranium deposition appears to have been controlled by braided channels and other vestiges of seismotectono-magmatic-belt activity (Tatsch, 1973a).

Bodovlje (Yugoslavia).

This deposit resembles that of Zirovski Vrh (q. v.).

Bohemia (Europe).

See Czechoslovakia.

Bolzano (Italy).

See Italian Sandstone.

Brive (France).

See France.

Campos Belos (Brazil).

This new deposit, about 400 km northeast of Brasilia, has indicated reserves of 1000 tons of ore assaying 0.12% to 5.0% U_3O_8.

Cardeville (Quebec).

This Soquem uranium deposit lies within the Lac Saint-Jean region about 300 km north of Montreal. The uranium content is roughly 0.05% U_3O_8. Besides uranium, the deposit also contains 0.22% Cb_2O_5, 0.09% TaO_5, and 0.14 ZrO_2.

Caucasus (U. S. S. R.).

See Vlasenchikhino.

Celje Basin (Yugoslavia).

These deposits resemble those of Zirovski Vrh (q. v.).

Cement (Oklahoma).

Hydrogeochemical anomalies have been used to delineate areas of potential uranium mineralization in south-central Oklahoma (See, e. g., Olmsted and Al-Shaieb, 1975). Many of the anomalies fall along vestiges of seismotectonomagmatic-belt activity associated with the axes of major oil-producing structrual trends which display bleaching of the Permian red beds of that area (Tatsch, 1974b). The bleaching appears to have been caused by the migration of hydrocarbons within paths of preferential flow associated with the vestiges of seismotectonomagmatic-belt activity in that area (Tatsch, 1972a). These hydrocarbons appear to have reduced the ferric iron to the more-soluble ferrous state, resulting in its removal by circulating ground waters. Specific ores deposited include uranium and calcium carbonate. The uranium mineralization is associated with a zone characterized by diagenetic carbonate, suggesting that the uranium was transported into the ore zone by ground waters moving upward along the flanks of the vestigal structure (Tatsch, 1973a).

Central Massif (France).

See France.

Chickasha (Oklahoma).

This southern Oklahoma deposit resembles that at Cement (q. v.).

Chinjra-Jari-Kashapat (India).

See Kishtwar.

Chirmatekri (India).

Significant concentrations of uranium within this part of the

Motur (q. v.) appear to have been controlled by permeable sandstones enclosed by clay beds.

Ciudad Rodrigo (Spain).

These uraniferous sandstones average better than 0.10% U_3O_8.

Colorado (U. S. A.).

These uraniferous marine black shales resemble those of Tennessee (q. v.).

Cottian Alps (Italy).

See Preit Valley.

Cox City (Oklahoma).

This southern Oklahoma deposit resembles that at Cement (q. v.).

Crocker Well (South Australia).

This deposit averages about 0.10% U_3O_8. Quartz monzonite intrudes metamorphic rocks and is intruded by alaskite, alaskite-pegmatite, and granitic pegmatite. The ore mineral is absite, a complex uranium-thorium titanite that averages about 32% UO_3 (Armstrong, 1974).

Czechoslovakia (Europe).

The uranium deposits of Czechoslovakia occur within the Permian sandstones within the intermontane basins of northern Bohemia and the Sudeten Mountains. The basins contain conglomerate, sandstone, and claystone, together with coal seams and organic matter. They appear to have been deposited under fluvial or lacustrine conditions. The uranium is concentrated in carbonate concretions and in bleached haloes within coal and dark shale. The enrichment probably originated post-depositionally by selective precipitation from mineralizing solutions, favored by a warm, dry climate, within these vestiges of seismotectonomagmatic-belt activity (Tatsch, 1973a). Other details of the uranium deposits of Czechoslovakia are contained in the literature (See, e. g., Pluskal, 1970).

Delta County (Colorado).

A vein-type source deposit containing U-238 and Th-232 appears to lie beneath the exposed Mesozoic sedimentary rocks of this area. This is suggested by the radioactivity of the spring flowing from the eastern part of the 30-km long spring-fed vestige of seismotectonomagmatic-belt activity along the Gunnison River (Tatsch, 1973a; Cadigan and Felmlee, 1975).

Dresden (Germany).

See Freital.

Eastern Alps (Austria).

The greenschist belts of the eastern Alps of Austria contain slighly metamorphosed uranium concentrations. These are associated mostly with sericitic quartzite and sericitic phyllite (See, e. g., Barthel, 1974).

Eola (Oklahoma)

This southern Oklahoma deposit resembles that at Cement (q. v.).

Erzgebirge (Czechoslovakia).

The Slovakian Erzgebirge mineralization was produced by Permian seismotectonomagmatic-belt activity from which the volcanism led to hydrothermal uranium deposition or to volcanogenic metal impregnations. The resulting mixed sedimentary-volcanic deposits formed uranium mineralization within vestiges of the Permian seismotectonomagmatic-belt in both Czechoslovakia and in northern Italy (Tatsch, 1973a).

The stratigraphic sequence begins with intercalations of sandstone, siltstone, and quartz porphyry. These are overlain by volcanoclastic beds, together with quartz porphyry, tuffite, sandy shale, and sandstone. These grade into greywacke, shale, marl, and evaporites.

Volcanic activity occurred primarily during the deposition of the middle and upper parts of the sequence. Within the middle part, a uranium-molybdenum mineralization developed; within the upper, a uranium-copper one. In both, there were some alterations of the autometasomatic type. The quartz porphyry is enriched in pitchblende, black oxides, molybdenite, and copper sulfides, plus some arsenopyrite, galena, and sphalerite.

Fieberbrunn (Austria).

See Chapter 7.

Figueira (Brazil).

During the Paleozoic and Mesozoic, several episodes of seismotectonomagmatic-belt activity in this area produced differential epeirogenic movements that formed the Ponta Grossa arch (Tatsch, 1973a). Beginning during the middle Paleozoic, this seismotectonomagmatic-belt activity continued intermittently until the late Mesozoic. The Devonian sediments were deposited during an actively subsiding phase of the seismotectonomagmatic-belt activity; the Permo-Carboniferous sediments, during a gently subsiding phase. The Permo-Carboniferous comprises epi-tillites , sandstone, and siltstones. The overlying Permian Rio Bonito formation comprises sandstone, siltstone, dark gray shale, and limestone. The basal part of this contains the uranium.

The uranium mineralization occurs as uraninite in sandstones within the interstices of the quartz grains. Uranium occurs also in organic complexes of siltstone, dark shale, and coal. Some secondary uranium occurs as uranocirite. The complex drainage system appears to correlate generally with the northwest-trending vestiges of seismotectonomagmatic-belt activity. The central and eastern parts of the Figueira area contain syngenetic uranium deposited in pelitic sediments in a swampy environment rich in organic material; the low-grade uranium here averages about 0.035%. The western part of the area contains epigenetic uranium deposits in mostly psammitic sediments in a fluvial or fluviodeltaic environment; this permitted minerlaizing solutions to percolate through the permeable conduits and to enrich these sediments to a average of about 0.2% U_3O_8.

Forstau (Austria).

See Austrian Alps.

France (Europe).

Some French uranium deposits are Jurassic (158 m. y.) but most are Permian (230 m. y.) (See, e. g., Barthel, 1974) All are concentrated about, and were probably derived from, the Massif Central. The most well known is Lodeve. Others include Brive, Laguepie-Monesties, Rodez, and St. Affrique.

Permian deposition began with Autunian (= Rotliegende = 270 m. y. ago) strata over a dissected, pre-Permain basement. A basal comglomerate is overlain by fine-grained sandstone with mud cracks, bituminous limestone, and floral shale. These are topped by the reddish and gray sandy and clayey deposits idenfified with Permian uranium in other parts of the world.

The depositional setting appears to have been swampy with rich flora. Some conifers occur within some areas that were temporarily flooded. Tropical, moist, hot climate was interspersed with some drier periods.

There are high-alkali, acid volcanic ash horizons. Mineralized horizons occur mainly on the flanks of channels, particularly at the interface between bleached (gray) sandstone and red sandstone. There are abundant flora and bitumens, with their carbonate content increasing upward.

The uranium minerals are carburan and some pitchblende. The lenticular, stratified mineralization zone appears to have formed under a combination of lithological and tectonic control, such as would be expected to have occurred within vestiges of late Paleozoic seismotectonomagmatic-belt activity. Enrichment appears to have been syngenetic. The source of the uranium appears to have been the weathering of granite within the Massif Central.

Within the depositional basins, the uranium was precipitated under reducing conditions. Migrating bitumens appear to have influenced the subsequent enrichment within joints and traps. There is a good correlation between the contents of uranium and carbon.

Besides uranium, other minerals enriched include Cu and Pb, plus some Zn enrichment along some joints. Accompanying metals include As, Ba, Mn, Mo, Sn, Sr, and V, plus some Co, Ni, and Sb. The uranium content falls inthe range 0.04% to 0.10% U_3O_8.

Other details are included in the literature (See, e. g., Barthel, 1974; Smirnov and Tugarinov, 1969).

<u>Franconia</u> (Germany).

See Stockheim.

<u>Freital</u> (Germany).

The Rotliegende coal of Freital, near Dresden, is enriched in uranium. The mineralization appears to have been syngenetic and to be related to the non-ferrous-metal shale in the sequence. The uranium content of the deposits ranges from 0.12% to 1.0% U_3O_8.

<u>Germany</u> (Europe).

Permian uranium mineralizations are found in both the lower and upper Rotliegende. Within the lower, several small concentrations exist in the Sarre and Nahe trough (See, e. g., Barthel, 1974). Near Obermoschel, six uraniferous horizons occur between sandstone and claystone sequences. The uranium is associated with carbonaceous material primarily (e. g., wood) and pyrite concentrations. It forms carburan, coffinite, and pitchblende, accompanied by chalcopyrite, marcasite, pyrite, and sphalerite. An epigenetic mineralization is suggested by the paucity of uranium in the claystone.

Some Rotliegende uranium deposits are related to the volcanic phase of Permian seismotectonomagmatic-belt activity (Tatsch, 1973a). In most cases, the sandstone is mineralized near the contact with rhyolitic rocks, or the sandstone contains conglomerate with mineralized volcanic pebbles.

Another type of mineralization is also found in some parts of the lower Rotliegende sandstone. This comprises uran-vanadium concentrations within spheroidal bodies, similar to those found in other parts of the world. The uranium content is within the range 0.005% to 0.3% U_3O_8. The dark gray to black cores are surrounded by bleached halos, probably resulting from the removal of the hematite pigment.

Other details are contained in the literarure (See, e. g., Barthel, 1974).

Glarus (Switzerland).

See Mürtschenalp.

Gorenja Vas (Czechoslovakia).

See Zirovski Vrh.

Granite Mountains (Wyoming).

These uraniferous deposits, which are peripheral to the Granite Mountains, appear to have derived their uranium from leaching of the granite in the mountains (See, e. g., Nkomo and Stuckless, 1975; Stuckless et al., 1975). Some of this leaching may have occurred at depths of over 700 m.

Guarda (Portugal).

These uraniferous sandstones average better than 0.10% U_3O_8. See also Serra da Estrada.

Healdton (Oklahoma).

This southern Oklahoma deposit resembles that at Cement (q. v.).

Helmsdale (Scotland).

The approximately 2500 km^2 of granites that were emplaced by the Caledonian episode of seismotectonomagmatic-belt activity contain scattered evidence of uranium mineralization. The minerals include some autunite, kasolite, and metatorbernite. These are associated with fractures into the Helmsdale granite that was intruded into Precambrian schist. The uraniferous anomalies in the stream sediments of this peat-covered area correlate with with the newer Caledonian granites. No significant deposit of unanium is known in this area.

Helvetian Nappe (Switzerland).

See Mürtschenalp.

Herault (France).

See Lodeve.

Idaho (U. S. A.).

The central part of Idaho contains some radioactive black minerals, mainly euxen te. These occur as accessory minerals in quartz monzonite and quartz diorite which are cut by a network of aplite and pegmatite dikes and stringers. Some euxinite occurs also in the pegmatite dikes. Mackin and Schmidt (1956) feel that the euxinite in the commercial placers at Big Meadow, in Bear Valley, were derived from a 13 km^2 area to the south. This suggests that the drainage area tributary to Big Meadow might be uraniferous.

Italian Metamorphic Rocks (Italy).

Some of the uranium of northern Italy has been concentrated into rocks of the greenschist facies. The epimetamorphic quartz-mica schist, sericitic phyllite, and sericitic quartzite appear to have been modified from greywacke and impure sandstone (See, e. g., Barthel, 1974). Very few traces now remain of the original relationship between the micro-crystalline uraninite and the floral relicts except within the lesser metamorphosed rocks. The same is true of the erstwhile relationship between the unaninite and graphite and phosphorite (apatite). This suggests that the marine ingressions were associated with the seismotectonomagmatic-belt episode that produced the metamorphism (Tatsch, 1973a). Although the age relations are ambiguous in some areas, it appears that the Alpine episode of seismotectonomagmatic-belt activity modified the Permian uranium deposits. See Preit Valley.

Italian Sandstones (Italy).

The Permian uranium deposits are mostly associated with the Verrucano province. These sandstone deposits lie within the facies areas that witnessed substantial seismotectonomagmatic-belt activity during the late Paleozoic (Tatsch, 1973a). This activity included the emplacement of granite and acid volcanic rocks penecontemporaneously with early Permian sedimentation that resulted in the accumulation of sedimentary-volcanic strata.

The major deposits (e. g., Val Daone and Val Rendena) are within areas of very thick (6 to 8 km) sequences. The mineralization occurred mainly at the interface between coarse-grained gray sandstone and fine-grained (Grödener) sandstone, reminiscent of similar occurrences in other parts of the world's late Paleozoic uranium deposits. Some lagoonal or lacustrine sediments are intercalated with deltaic or fluviatile elements. The sorting is generally poor with load casts, cross-bedding, and ripple marks.

Enrichments are generally concordant with stratification, but there are some lenticular mineralizations. These include some carbonized wood. Besides uranium, the mineralization includes pyrite plus some chalcopyrite, galena, sphalerite, and tetrahedrite. The uraninite deposits were apparently laid down by circulating solutions that extracted the uranium from volcanic rocks (See, e. g., Mirrempergher, 1970, 1974) either finely distributed or as hydrothermal enrichments.

Mineralization within the alpino-type zones appears to have been epigenetic (See, e. g., Barthel, 1974); that within basins with restricted sedimentation appears to have been a synsedimentary uranium enrichment in epicontinental pelite-siltite under the influence of pyrite and organic material.

Near Bolzano, the uranium occurs with galena and forms a syngenetic, stratiform, red-bed association. Most of the lenticular enrichments are in floral medium-grained gray sandstone. Both uraninite and colloform pitchblende are present. The pitchblende appears to have been mobilized by seismotectonomagmatic-belt activity, primarily of a tectonic nature. The uranium content, ranging from 0.05% to 0.8%, appears to have originated from volcanic rocks associated with vestiges of seismotectonomagmatic-belt activity (Tatsch, 1973a).

Other details are contained in the literature (See, e. g., Bondi et al., 1973; Tarthel, 1974).

Jari (India).

See Kishtwar.

Jatoba (Brazil).

Some low-grade (0.025% U_3O_8) uranium is associated with the clay and phosphatic matrices of sandstone deposited during the Devonian coastal environments of northeastern Brazil (DeAndrade-Ramos and Fraenkel, 1974).

Kaoleri (India).

Significant concentrations of uranium within this Motur (q. v.) bed appear to have been controlled by permeable sandstones enclosed by clay beds. No roll-type controls have been observed in this area.

Kara-Tau Mountains (Kazakhstan).

This deposit is similar to that at Phosphoria (q. v.) except that the U. S. S. R. deposits are Cambrian rather than Permian. See also USSR Black Shales, in Chapter 9.

Kashapat (India).

See Kishtwar.

Kentucky (U. S. A.).

These uraniferous marine black shales resemble those of Tennessee (q. v.0.

Kiroli (India).

See Kishtwar.

Kishtwar (India).

Veins of pitchblende occur in a number of areas along tension joints in the crests of anticlinal folds or in a quartzite of the Paleozoic Jaunsar sequence of India. This sequence forms a fairly persistent horizon in those vestiges of seismotectonomagmatic-belt activity that are referred to as the Lesser Himalayan tectonic province (Tatsch, 1972a). These vestiges are traceable from the Doda district of Jammu-Kahmir state in the northwest to the Nepal border in the southeast. Clusters of these veins with variable vein-density have been located in Kishtwar, Jammu-Kashmir (See, e. g., Udas and Mahadevan, 1974). The geological setting of this area and the causal vestiges of seismotectonomagmatic-belt activity are similar to those of the Chinjra-Jari-Kashapat area of Himachal Pradesh. The quartzites are overlain by chlorite schists and gneisses, all now folded and prsserved in syncline vestiges of subsequent seismotectonomagmatic-belt activity (Tatsch, 1973a).

The pitchblende veins range in thickness from a few mm to a few cm and are as much as 10 m long. They rarely extend deeper than 5 m. Some pitchblende also occurs as intergranular fillings in the quartzite along the vein margins. In Kiroli (Pathoragarh) and other areas, the veins occur along sheared fractures. Associated with the pitchblende are minor amounts of calcocite, chalcopyrite, and pyrite. Secondary uranium mineralizations include beta-uranophane, uranogummite (ocher) and vandendriesscheite.

It appears that the mineralizing solutions for these pitchblende veins were magmatogenic-hydrothermal vestiges of seismtoctonomagmatic-belt activity. The immediate vestige was probably granite (See, e. g., Udas and Mahadevan, 1974). Similar uraniferous vestiges are found in the Simla area af the Central Himalayan gneissic complex. In both these cases, as well as in similar cases, an alternative scenario uses circulating ground waters to produce the veins from the uraniferous vestiges of seismotectonomagmatic-belt activity. In this scenario, an added phase transfers the uranium from the overlying chlorite schists and some gneisses into the circulating ground waters. These then deposit the pitchblende veins. In either case, the uranium vestiges of seismotectonomagmatic-belt activity supply the uranium.

Kolm (Sweden).

This shale deposit, near Ranstad, averages about 0.03% U_3O_8. The mode of deposition appears to have been low-temperature adsorption onto organic or phosphatic material in organic shales (See, e. g., Barnes and Ruzicka, 1972).

Kupferschiefer Deposits (Germany).

The Kupfershiefer is a transgressive marine deposit at the base of the upper Permian Zechstein limestone. This was deposited within a 3000 km long and 300 km wide vestige of earlier seismotectonomagmatic-belt activity (Tatsch, 1975a). The sequence normally comprises too components: (1) a conglomerate of fine-grained, light-gray sandstone (Weissliegende); and (2) an overlying bituminous Kupfershiefer about 1 m thick. This consists of very bituminous (15% bitumen), bituminous clay-marl, slightly bituminous marl, and dolomite (sometimes sandy) limestone. It is characterized by a horizontal and vertical zoning within the metal distribution. The richest copper content (up to 3%) is found in the coastal regions. Away from the coast, the metals are mainly lead (0.5%) and zinc (2%).

Within the Mansfeld region, this distribution of metals is particularly noticeable. Copper was deposited in the lower, argillaceous levels (Tatsch, 1975a). Overlying this is a filling of plumbeous (lead-bearing) calcareous shells and some zinc encrustations of these shells. Silver appears to have been deposited with the copper; cobalt and nickel, with the carbonaceous minerals; and molybdenum and vanadium with the clay minerals. The uranium appears to have been precipitated early, in association with a polymerized bituminous, pitchy horizon in the lower part of the Kupferschiefer.

The uranium content of the Kupfershiefer ranges from about 0.002% to 0.05%. A synsedimentary or syndiagenetic origin for the uranium (and for the non-ferrous metals) is suggested by (1) their uniform precipitation over a broad area, and (2) their zoning dependence on both the solubility product and the paleogeography.

Some epigenetic enrichments occur along vein fissures (See, e. g., Barthel, 1974). In these cases, a paragenetic association of Bi-Co-Ni with barite and a Mo-U-Re enrichment was followed by a Cu mineralization rich in Ag and Ge. Some of this paragenesis appears to have been dissolved by migrating bitumen followed by redeposition. Within the Mansfeld area, some pebbles and propylitized volcanic rocks in the Weissliegende and the underlying Permian have undergone a Cu-Po-Zn mineralization.

Metal deposits comparable to the Kupfershhiefer are found in the bituminous shale and the intermontane basins, e. g., the Sarre-Nahe (q. v.)

Kyuenelekeen (U. S. S. R.).

See Anabar.

Laguepie-Monesties (France).

See France.

Laguna (New Mexico).

See Grants.

Lisbon (Utah).

This deposit is associated with vestiges of late Paleozoic and Laramide episodes of seismotectonomagmatic-belt activity (Tatsch, 1973a). It is discussed in Chapter 7.

Lodeve (France).

The Herault area of France contains areas of retarded sedimentation which provided a metallotect for uranium. The carrier beds of this stratiform regional anomaly contain uranium plus some astatine, lead, and molybdenum (See, e. g., Herbosch, 1974). These beds are mostly finely laminated and bituminous pelite. Lacustrine episodes occurred during the lower Permian within a thick continental pelite-sandstone fill of red beds. Besides the heavy metals, there is some finely dispersed organic material reminiscent of similar deposits in other parts of the Eurasian Permian.

In some respects, the Lodeve basin resembles the Eocene Green River oil shales of the western U. S. A.; but in others, it is quite different. The uranium could have been from the Massif Central that lies to the north.

Lombardy Trough (Italy).

See Italian Sandstone.

Lukachukai (Arizona).

This deposit lies at extensions from the Uravan - Slick Rock (q. v.) and Grants (q. v.) vestiges of seismotectonomagmatic-belt activity (Tatsch, 1973a).

Madaoula (Niger).

These peneconcordant uranium ores in sandstone contain over 6000 tons of ore averaging better than 0.10%.U_3O_8.

Madhya Pradesh (India).

These Permo-Carboniferous lenses contain 0.03% to 0.04% U_3O_8. The mineralization occurs within argillaceous lacustrine and fluvial sandstones. These have been eroded from Precambrian gneissic and granitic vestiges of seismotectonomagmatic-belt activity (Tatsch, 1973a).

Mansfeld (Germany).

See Kupferschiefer.

Mansingpura (India).

Significant concentrations of uranium occur within this Motur (q. v.) bed. The uranium deposition appears to have been influenced by the interfingering of clay beds. These, in turn, are identifiable with early Paleozoic vestiges of seismotectonomagmatic-belt activity (Tatsch, 1973a).

Massif Central (France).

See France.

Mayrhofen (Austria).

See Austrian Alps.

Mecsek Mountains (Hungary).

Uranium enrichments occupy roughly 500 m of the upper Permian fluviatile sandstone of the Mecsek Mountains. The deposits comprise sandstone and siltstone with siliceous and calcareous cement plus some lenticular claystone intercalations. These are mainly gray and are overlain by red sandstone with iron-rich argillaceous and calcareous cement (See, e. g., Tatsch, 1975g).

The uranium occurs along the borders of swampy-to-fluviatile deposits associated with floral matter and arkose. The material appears to have eroded from granitic rocks enriched in Bi, Co, and Ni. Simultaneously, some basic igneous detritus was deposited. The uranium content ranges from 0.008% to 0.015% U_3O_8. The uranium mineralization was accompanied by the formation of the sulfides of Co, Cu, Ni, Pb, and Zn.

In many respects, the Mecsek deposits are similar to the uraniferous deposits of other parts of the world that are now associated with Paleozoic and Mesozoic vestiges of seismotectonomagmatic-belt activity (Tatsch, 1973a). These include particularly those of Konigstein (Germany), Hamr (Czechoslovakia), and Uchkuduk,

Sa-byr-Say (U. S. S. R.). The mineralization is confined to fluviatile, deltaic, littoral marine, and lacustrine sediments controlled by tectonic traps associated with vestiges of seismotectonomagmatic-belt activity of that area (Tatsch, 1973a).

Mittelberg (Austria).

The uraniferous mineralization of this Salzburg copper deposit appears to have been derived by metamorphic mobilization from sedimentary concentrations within the Permian sandy shale of the greywacke zone (See, e. g., Siegl, 1972).

Mongollon (Arizona).

The uraniferous sandstones exposed along the Mongollon Rim between Oak Creek Canyon and the Fort Apache Indian Reservation in eastern Arizona are being tested to determine the source of the uranium in the coarse-grained rocks of the area (See, e. g., Anon., 1975b). Initial analyses suggest that this uranium is traceable to the same vestiges of seismotectonomagmatic-belt activity that are responsible for other sandstone uranium deposits in that part of the U. S. A. (Tatsch, 1973a).

Montana (U. S. A.).

These uraniferous marine black shales resemble those of Tennessee (q. v.).

Motur (India).

The Motur beds comprise fluvial deposits of current-bedded immature gritty greenish and gray feldspathic sandstones, clays, and shales, plus lenses of conglomerates and pebble beds. The pebbles comprise massive quartz, chert, jasper, and granite. There is also some anthracite coal. The moderately northerly dipping (5° to 20°) beds are locally folded into broad anticlines and synclines.

The uraniferous occurrences are controlled by bedding. No roll-type mineralization has been observed (Udas and Mahadevan, 1974). The concentration appears to have been formed by the remobilization and enrichment appears to have resulted from the circulation of ground waters. The mineralization occurs as black oxides, as coffinite, and as adsorbed elements of clays. The concentration ranges from a trace to 0.20% U_3O_8. The uranium appears to have been introduced from sources other than the overlying Bijori formation. Little or no thorium occurs within the Motur uranium deposits.

Motur beds with significant concentrations of uranium include those at Bodipani, Chirmatekri, Kaoleri, Mansingpura, and Polapathar.

Mühlbach (Austria).

A rather unique uranium deposit occurs in association with the well-known hydrothermal copper vein of Mühlbach, near Bischofssafen. The copper vein cuts black siluric schist, which is overalin by violet quartzites of lower Permian age. These quartzites contain small stratiform layers of fine-grained brannerite and uraninite. On both sides of the vein, there are some black nodules of uraninite. These appear to have resulted by leaching and redeposition of the sedimentary ore in the quartzite (See, e. g., Siegl, 1972). Uniquely, some of the uraninite nodules contain native gold in their cracks (See, e. g., Petrascheck et al., 1974; Tatsch, 1975f). The gold appears to have been precipitated from a very weak solution, by the reduction effect of the tetravalent uranium upon its oxidation to the hexavalent form.

Mussoorie (India).

Significant concentrations of uranium occur within the phosphorites of the Mussoorie syncline within the lower Himalayan district of Dehra Dun, Uttar Pradesh. The phosphorites outcrop along a periphery of doubly-plunging suncline for a distance of about 120 km (Udas and Mahadevan, 1974), striking roughly northwest-southeast.

The phosphorites lie between the underlying Krol limestone and the overlying Tal shales and sandstones. The phosphatic rocks are dull gray to brownish in color. They may be granular, laminated, nodular, or platy.

Significant concentrations of uranium are confined to the area between Mussoorie and Sahasradhara at the southrrn limb of the Mussoorie syncline. The uranium averages about 0.03% U_3O_8. The overlying black shales contain about 0.01% to 0.02% U_3O_8. There does not seem to be a direct correlation between the uranium and the P_2O_5 concentrations.

When the geological features of the unaniferous and non-uraniferous phosphates of India are compared, it is found that the non-uraniferous phosphates contain stromatolitic structures (e. g., Jhamarkorta (Udaipur, Rajasthan) and Pithoragarh, U. P.). This suggests that the algal and bacterial life that produced the stromatolitic structures thrives only under conditions having oxygen; such an environment does not support the required redox potentials required to precipitate uranium (Udas and Mahadevan, 1974).

Compared with the Permian phosphorites of the U. S. A. (See, e. g., Phosphoria), the uraniferous phosphorites of Mussoorie: (1) have higher enrichments of Ba, Co, Mo, Ni, and V; (2) have lower enrichments of Cr, La, and Sc; and (3) have about equal enrichments of Sr, Th, Y, and Zr. The Mo and Ni enrichments may reach roughly 0.2% over significant strike lengths. These enrichments, plus the association with black shales, confirms the extreme Eh conditions that favor uranium enrichment.

Nacimiento-Jemez (New Mexico).

The deposits in this area, which span the Permian-to-Tertiary age, are discussed under this heading in Chapter 7.

Närke (Sweden).

These uraniferous marine black shales resemble those in Tennessee (q. v.) except that the ores average about 0.03% U_3O_8. See also Sherman (1972).

Navaho (=Navajo) (Arizona).

The Navaho and Hopi Indian Reservations contain some uranium deposits within or associated with roughly 250 diatremes. Some of these diatremes resemble the diamond pipes of South Africa. They contain pyroclastic material similar to kimberlite. When fully developed, these diatremes contain bedded tuff and limestone, plus some laminated clay and siltstone, evaporites, and some bedded chert, in the upper portions. In the lower portions, these sediments grade into more-massive tuff, breccia, large blocks of country rock, agglomerate, and solid igneous rock. The volcanic rocks associated with these intrusive vestiges of seismotectonomagmatic-belt activity are nearly all alkaline basalts. These resemble the Colorado Plateau olivine basalts. The uranium concentration is only about 0.002%. The uranium is present in extremely fine-grained form. Minerals identified include bebigite, carnotite, and metatyuyamunite, plus some cobalt, lead, nickel, silver, and thallium. The uranium appears to have originated syngenetically from hydrous, phosphate-rich magma in the lower part of the diatremes.

North Carolina (U. S. A.).

The marine phosphorites in North Carolina contain about 5 million tons of ore averaging 0.006% to 0.012% U_3O_8.

Novazza (Italy).

This deposit, northeast of Bergamo, occurs in Permian sediments comprising lavas, tuffs, and shales. The main mineralization of pitchblende occurs within fractured porous volcanic rocks. The average grade at Novazza is about 0.15% U_3O_8.

Novazza - Val Seriana (Italy).

This new deposit in the Alps is scheduled to go on stream in 1979. Roughly 2.5 million tons of ore grade about 0.1% U_3O_8.

Oklahoma (U. S. A.).

These uraniferous marine black shales resemble those of Tennessee (q. v.).

Oslo (Norway).

The uraniferous upper Cambrian Olenid series of the Oslo region is a very monotonous sequence of black carbonaceous shales (alum shales) with nodules and layers of black, bituminous limestone (See, e. g., Strand and Kulling, 1972). Besides uranium (0.017%), the black alum shales are rich in iron sulfide and 15% to 17% carbon (Tatsch, 1975g). Unlike the alum shales of middle Sweden, the Norwegian alum shales do not yield oil and gas by distillation (See, e. g., Tatsch, 1974b).

Pachmarhi (India).

These fluvial sediments contain some uranium and thorium, but not as much as do the Bijori (q. v.) deposits. The beds comprise white sandstones with quartz pebbles, feldspar fragments, and reddish clays (Udas and Mahadevan, 1974). The concentrations are mostly monzonite and zircon.

Panhandle (Texas).

The red beds of the Texas Panhandle area are being tested to determine the origin, evolution, and present characteristics of their uraniferous contents (See, e. g., Anon., 1975b). Initial analyses suggest that this uranium is traceable to the same Proterozoic and early Phanerozoic vestiges of seismotectonomagmatic-belt activity that account for red-bed-associated uraniferous deposits in other parts of the world (Tatsch, 1973a).

Parana Basin (Brazil).

This basin of southern Brazil compirses Devonian-to-Cretaceous sediments covering over one million km^2. Cretaceous basalts covered 80% of the area during a late Mesozoic episode of seismotectonomagmatic-belt activity (Tatsch, 1973a). The uraniferous areas appear to be associated with areas near the contacts with granite. This is particularly noticeable in the Permo-Carboniferous sediments (See, e. g., deAndrade-Ramos and Fraenkel, 1974). One of the important deposits is at Figueira (q. v.) in the coal-producing region of Rio do Peixe.

Peredovoyi (U. S. S. R.).

See Vlasenchikhino.

Phosphoria (U. S. A.).

The Permian Phosphoria formation deposits underlie about 400,000 km^2 of Idaho, Montana, Utah, and Wyoming (McKelvey and Carswell, 1956). The uranium content ranges from 0.001% to 0.075% U_3O_8. The richest beds are those that are thicker than one meter and containing over 30% P_2O_5; these evarage 0.012% to 0.024% U_3O_8. There is a positive correlation between the uranium and phosphate contents. Other details are in the literature (See, e. g., Maughan, 1975).

Piaui-Maranhao (Brazil).

This is part of the low-grade (0.025% U_3O_8) that is associated with the clay and phosphatic matrices of sandstones deposited during the Devonian coastal environments of northeastern Brazil (De

Andrade-Ramos and Fraenkel, 1974).

Pithoragarh (India).

See Kishtwar.

Pluto Bay (Saskatchewan).

This northern Saskatchewan deposit is within granite greiss immediately underlying a molybdenum zone. The mineralization zone extends roughly 30 km around the rim of a shallow syncline. The uranium content is about 0.05% U_3O_8.

Poland (Europe).

The few uranium deposits in southern Poland, north of the Sudeten Mountains, resemble those of Czechoslovakia (q. v.).

Polapathar (India).

Significant concentrations of uranium occur within this Motur (q. v.) bed. The uranium deposition appears to have been influenced by the differences in the permeability of the sandstones within this vestige of seismotectonomagmatic-belt activity (Tatsch, 1973a).

Portugal (Europe).

Portugal's uranium deposits, which span the Precambrian to the Mesozoic, are discussed in Chapter 7.

Preit Valley (Italy).

Some of the Permian uraninite of the Preit Valley of the Cottian Alps appears to have been metamorphosed by the Alpine episode of seismotectonomagmatic-belt activity (Tatsch, 1973a). The unmetamorphosed rocks appear to have been clastic sediments, now metamorphosed to sericite-chloritic schist. The original (Permian) uranium was associated with pyrite and floral remains. The Alpine metamorphism mobilized the uranium and associated sulfides. These formed into lenticular and nodular enrichments. Some cataclastic deformations are now evident within the pitchblende and pyrite. Softer ores (e. g., sphalerite) display some myrmekitic (wart-like) intergrowths. These myrmekites appear to have formed in conjunction with the plutonic phase of the Alpine seismotectonomagmatic-belt activity.

Radece (Yugoslavia).

These deposits resemble those of Zirovski Vrh (q. v.).

Ranstad (Sweden).

See Billingen.

Rodez (France).

See France.

Romania (Europe).

The uranium deposits of Romania accur within a series of multi-colored rocks representing a transitional stage between gray and red rocks (See, e. g., Kornechuk and Burtek, 1974). In this transition, the ore is associated with gray and gray-green rocks comprising conglomerates, greenstones, gritstones, and sandstones. The rocks display good permeability and sharp lithological contrast. They were formed in alluvaal and lacustrine facies. The accumulation appears to have been controlled by vestiges of seismotectonomagmatic-belt activity (Tatsch, 1973a). These vestigal controls included lithological, structural, and tectonic screening. Most ore bodies, which are stratified and lenticular, lie concordantly with the strata. The mineral is pitchblende, finely dispersed with bitumen in some cases, plus some uranium black.

The genesis was complex. In particular, the bitumen-associated deposits appear to be related to strongly metamorphosed varieties of anthraxolites of the petroleum series (Tatsch, 1974b). The petroleum appears to have migrated and concentrated into porous and fissured rocks, followed by metamorphic phases of subsequent seismotectonomagmatic-belt activity. Many of the source rocks are dark, organic-rich intercalated sandstones and siltstones belonging to the upper Carboniferous and other upper Paleozoic horizons. In this respect, these Romanian deposits resemble the organic material in the Grants (q. v.) type primary ores.

The highest state of oxidation occurs within the bitumens containing the highest concentrations of uranium. Other deposits appear to be associated with other minerals. There has been a reduction in the amount of iron present as oxides (See, e. g., Tatsch, 1975g).

Rotliegende (Germany).

See Kupferschiefer.

Russia (Eurasia)

See USSR.

Saint Affrique (France).

See France.

Saint Bernhard (Switzerland).

The St. Bernhard nappe in the Canton of Valais contains a metamorphically altered U-Cu-Mo-Pb mineralization zone. The deposit is associated with chloritic-sericitic schist and sericitic-albitic gneiss. The original rocks appear to have been a sequence of upper Carboniferous to late Triassic sedimentary rocks; the mineralized horizons, lower Permian. The original uranium and nonferrous metal mineralizations appear to have been sedimentary.

During the metamophism associated with Alpine seimmotectonomagmatic-belt activity, the minerals were mobilized and underwent complex recrystallization (See, e. g., Tatsch, 1973a). This mobilization caused some of the ores to migrate to the neighboring rocks.

Sarre-Nahe (Germany).

The metal deposits of the Sarre-Nahe depression of the Rotliegende resemble those of the Kupferschiefer (q. v.). Basically, the Sarre-Nahe deposits are contained in the bituminous shale of some intermontane basins of that area. Within a 3-to-5-cm section of bituminous shale in this depression, there appears to have been a seasonal change between deposition of very calcareous and slightly bituminous laminations, on the one hand, and very bituminous and slightly calcareous horizons on the other.

Within the very bituminous material, the uranium is enriched to 0.005% U_3O_8. Other metals include concentrations up to 0.004% Co, 0.02% Cr, 0.05% Cu, 0.01% Mo, 0.015% Ni, 0.01% Pb, 0.035% V, and 0.05% Zn. These enrichments appear to have been syngenetic.

Much of the coal within the Sarre-Nahe (and other Rotliegende) depressions is enriched in uranium. The content ranges from 0.002% to 0.04% U_3O_8. Most of this uranium is adsorbed on the coal and on the adjacent shale. Other Rotliegende depressions that resemble Sare-Nahe include Freital (q. v.) and Stockheim (q. v.).

Sava Folds (Yugoslavia).

The widest extent of the Gröden beds in the northwestern part of Yugoslavia is within the Sava Folds. These regions comprise a 130-km by 24-km relict of seismotectonomagmatic-belt activity that produced an area of complex tectonics at the juncture of the Alps and the Dinarides (Tatsch, 1972a). Some of this area was modified by later episo es of Alpine seismotectonomagmatic-belt activity (See, e. g., Lukacs and Florjanicic, 1974).

Schladming (Austria).

The Schladming deposit, within the lower Tavern, is enriched in Permian sericitic quartzite and sericitic phyllite. Some observers feel that the processes that formed this deposit were similar to those that enriched some Permian sandstones. Others favor a mechanism similar to that which mineralized the greenschist facies in the Alps (See, e. g., Barthel, 1974). According to the Tectonospheric Earth Model, these deposits are vestiges of Permian and older episodes of seismotectonomagmatic-belt activity that occurred in that area (Tatsch, 1973a).

Simla (India).

See Kishtwar.

Skofja Loka (Yugoslavia).

These deposits resemble those of Zirovski Vrh (q. v.).

Stockheim (Germany).

Some of the coal seams of the Stockheim basin in the lower Franconia contain significant amounts of uranium. This mineralization has affected both the coal and the associated shale of that area. Close scrutiny suggests that the uranium is related to the coal seams (See, e. g., Barthels, 1974). These seams, in turn, are related to vestiges of late Paleozoic seismotectonomagmatic-belt activity in that area (Tatsch, 1973a).

Similar deposits occur in Freital (q. v.) and Sarre-Nahe (q. v.).

Sudeten (Czechoslovakia).

See Czechoslovakia.

Tennessee (U. S. A.).

These uraniferous marine black shales, rich in organic matter, were probably deposited under anaerobic conditions within shallow-water epicontinental seas existing there during the late Devonian and early Mississippian (See, e. g., Swanson, 1961). These deposits are associated with vestiges of seismotectonomagmatic-belt activity that occurred there during the early and middle Paleozoic. Occurring within the upper member of the Chattanooga shale, these deposits are 3 to 4 m thick and cover an area of roughly 10,000 km^2. They average about 0.07% U_3O_8.

Associated with this deposit are those of Alabama, Kentucky, Arkansas, Texas, Oklahoma, and possibly those extending from Texas to Montana, comprising an area of roughly 2 million km^2, The uraniferous strata are about 10 m thick and average about 0.0035% U_3O_8.

<u>Texas Paleozoic</u> (Texas).

Very little is known about the Paleozoic uranium deposits of Texas. But they appear to be associated with the same vestiges of seismotectonomagmatic-belt activity that were reactivated during later episodes of seismotectonomagmatic-belt activity (See Texas Mesozoic, in Chapter 7).

<u>Trentino</u> (Italy).

These deposits occur in Permian sandstones and in ignimbrites. The grade is about 0.1% U_3O_8.

<u>Trzic</u> (Yugoslavia).

These deposits resemble those of Zirovski Vrh (q. v.).

<u>Tuscaloosa</u> (Alabama).

The uraniferous sandstones of the Tuscaloosa group, in central Alabama, are being tested (See, e. g., Anon., 1975b). The uranium in the Pottsville formation of this area appears to be traceable to the same vestiges of seismotectonomagatic-belt aciivity that are responsible for other sandstone uranium deposits in that part of the U. S. A. (Tatsch, 1973a).

Tyuya Myuyum (U. S. S. R.).

These Permo-Carboniferous and Tertiary deposits occur deep within a basin of Ferghana Valley (q. v.). The uranium content, which grades to as much as 1% U_3O_8, depends on the concentration of sodium sulfate, sodium chloride, and other salts. The concentrations of uranium are in the dry remnant encrustations of these salts. They originated as fairly low-temperatuee chemical precipitations. See also Ferghana.

Udrichhappar (India).

See Bartalao.

Urupskoye (U. S. S. R.).

See Vlasenchikhino.

USSR Alkaline Subvolcanics (U. S. S. R.).

These uraniferous deposits occur within Paleozoic anticlinal vestiges of seismotectonomagmatic-belt activity in various parts of the U. S. S. R. that were affected by Caledonian episodes of seismotectonomagmatic-belt activity (Tatsch, 1973a). The mineralization is in rocks of an alkaline subvolcanic complex association represented by porphyritic rocks intruded by Silurian nepheline syenite and Devonian monzonite, nordmarkite, and alaskite granite. There are two mineral associations containing pitchblende: (1) pitchblende-cyrtolite, and (2) pitchblende-chalcopyrite.

USSR Caledonian Slates (U. S. S. R.).

The uraniferous deposits in these carbonaceous-cherty slates ocuur within Paleozoic anticlinal vestiges of seismotectonomagmatic-belt activity in various parts of the U. S. S. R. that were affected by Caledonian episodes of seismotectonomagmatic-belt activity (Tatsch 1973a). The uraniferous deposits occur within slightly metamorphosed folded sediments of vestiges of Ordovician, Silurian, and Devonian age. The vestigal anticlinal structures are cut by a fault zone that appears to have controlled the granitoid intrusions and vein-type uranium mineralization associated with this seismotectonomagmatic-belt activity. The Ordovician deposits are concordant; the later ones, discordant.

USSR Uraniferous Deposits of the Middle and Late Paleozoic (U. S. S. R.).

The middle and late Paleozoic vestiges of seismotectonomagmatic-belt activity in the U. S. S. R. contain numerous uranium-molybdenum mineralizations (Tatsch, 1973a; Modnikov et al., 1971). These mineralizations lie mostly in a transition zone between episodes of Caledonian and Hercynian seismotectonomagmatic-belt activity. The uranium-molybdenum ore deposit is located near the fractures associated with the northwest-trending vestiges within a subvolcanic body of felsite-porphyry of the Devonian series. The main ore-forming minerals are pitchblende, molybdenite, and fluorite. Accessory minerals are galena, hematite, and pyrite (See, e. g., Tatsch, 1975g). The pitchblende is associated with micas, kaolinite, barite, montmorillonite, and related minerals reminiscent of the associations seen in numerous middle and late Paleozoic uraniferous deposits in other parts of the world (Tatsch, 1973a). Some antimony, arsenic, silver, and tellurium also occur with the pitchblende.

USSR Uraniferous Pyrites (U. S. S. R.).

Several vestiges of Ordovician seismotectonomagmatic-belt activity in the U. S. S. R. contain uraniferous pyrites (Tatsch, 1973a; Komarov and Yergorov, 1970). These deposits occur in weakly metamorphosed vestiges of seismotectonomagmatic-belt activity, usually in veins of three generations: (1) pyrite older than uranium, (2) pyrite and uranium of the same age, and (3) uranium older than pyrite (See also Tatsch, 1975g). The uranium content ranges from traces to 0.002% U_3O_8.

Valais (Switzerland).

See St. Bernhard.

Val Daone (Italy).

See Italian Sandstones.

Val Pescara (Italy).

See Italian Sandstones.

Val Rendana (Italy).

See Italian Sandstones.

Västergotland (Sweden).

These upper Cambrian uraniferous marine black shales resemble those in Tennessee (q. v.), except that the ores overage about 0.003% U_3O_8. See also Sherman (1972). Similar deposits occur at Närke.

Velma-Cruce (Oklahoma).

This southern Oklahoma deposit resembles that at Cement (q. v.).

Verrucano (Italy).

See Italian Sandstones; Mürtschenalp.

Viseu (Portugal).

These uraniferous sandstones average better than 0.10% U_3O_8.

Vlasenchikhino (U. S. S. R.).

This deposit is part of the Urupskoye ore field in the northwestern part of the middle Paleozoic zone of the Peredovoyi range in the north Caucasus. Uraniferous pyritiferous deposits are localized in the Devonian series of seismotectonomagmatic-belt activity (Tatsch, 1973a, 1975g; Bogush and Savchenko, 1971). The ore occurs in lens-shaped strata reminiscent of similar deposits in the middle Paleozoic vestiges of seismotectonomagmatic-belt activity in other parts of the world (Tatsch, 1973a).

Vosges (France).

The black shales on the eastern slopes of the Vosges mountains contain uranium in the form of organo-uranium complexes. These deposits, which grade about 0.08% U_3O_8, are quite extensive and are reminiscent of the low-grade Chattanooga shales (See Tennessee), the Swedish schists, and similar deposits in other parts of the world. They seem to be associated with vestiges of an episode of late Paleozoic seismotectonomagmatic-belt activity (Tatsch, 1973a).

Weissliegende (Germany).

See Kupferschiefer.

Wyoming (U. S. A.).

These uraniferous marine black shales resemble those of Tennessee (q. v.).

Yugoslavia (Europe).

Some uraniferous veins and peneconcordant sandstones of Yugoslavia average better than 0.10% U_3O_8. See also Zirovski Vrh.

Zambesi (Zambia).

Uranophane occurs within sandstone of the Karoo system near

the Zambesi rift valley.

Zivorski Vrh (Yugoslavia).

This Slovenian deposit, within the lower Permian Gröden sandstone, is concentrated in lenses within fluviatile sandstone. In this area, continental sediments extend from the upper Carboniferous to the upper Permian. The lower strata comprise almost a km of gray-green sandstone intercalated with red siltstone. Above is red sandstone with gray intercalations. The gray-to-red color change parallels bedding.

Uranium occurs within lenses of the lower gray sandstone, restricted to a 150-km sequence within pitchblende occuringg as a cement or finely-distributed material in organic matter. Some other minerals occur independently of the uranium, e. g., copper sulfides. Some rare galena and chalcopyrite associations within uranium occur. The uranium content ranges from 0.05% to 5.0% U_3O_8.

The composition of the sediments suggests the presence of volcanogenic detritus. Some tuffitic intercalations are observed. The fact that the uranium in the gray sandstone is more soluble than that in the red sandstone suggests why the uranium is more highly concentrated in the gray sandstone. The uranium is associated with vestiges of middle and late Paleozoic episodes of seismotectonomagmatic-belt activity (Tatsch, 1973a).

Other details of this deposit are contained in the literature (See, e. g., Lukacs and Florjancic, 1974; Samana, 1973; Markov and Ristic, 1974).

Conclusions Regarding the Paleozoic Uranium Deposits.

Not all uranium deposits that were emplaced during the Paleozoic are known. Some may never be discovered; others will be found through the use of advanced geological concepts and better exploration techniques. One area that probably contains unknown but discoverable uranium deposits lies along the Gibraltar-to-Bengal quadrispherical arc, particularly within the Alpine geosyncline (See, e. g., Tatsch, 1973a; Evans, 1975). An extension of this same principle suggests that the same situation pertains to the segments of the other 11 quadrispherical arcs that were active during the Paleozoic episodes of seismotectonomagmatic-belt activity.(Tatsch, 1973a). Close scrutiny of the known Paleozoic uranium deposits reveals that almost all of them appear to be associated with those areas of hhe Earth's surface where vestiges of Paleozoic seismoteconomagmatic-belt activity remain today.

Chapter 9

PROTEROZOIC SEISMOTECTONOMAGMATIC BELTS AND THE ASSOCIATED URANIUM DEPOSITS

The uranium deposits associated with the Proterozoic seismotectonomagmatic belts comprise, according to the Tectonospheric Earth Model concept, part of the Earth's "original" early-Archean uranium that was "reworked" by subsequent seismotectonomagmatic-belt activity during the Archean and the Proterozoic but not during the Phanerozoic. Deposits of the Phanerozoic have been considered in earlier chapters. It is not too surprising that the Earth's Proterozoic uranium deposits are found primarily within the platforms and their folded peripheries, including those of western Siberia, southern and western Africa, the United States, Canada, Brazil, India, Scandinavia, and Australia. These are the main areas where vestiges of Proterozoic seismotectonomagmatic-belt activity still remain today.

In analysing the distribution patterns of the Proterozoic uranium deposits, it is well to keep two things in mind: (1) the the continental areas were not necessarily the same during the Proterozoic as they are today (See, e. g., Symons, 1975); and (2) the Proterozoic lasted for 2 b. y., or over three times as long as the entire Phanerozoic and about 30 times as long as the entire Cenozoic. This, considered against the background of continuous seismotectonomagmatic-belt activity in various parts of the Earth, makes it meaningless to identify _single_ Proterozoic patterns of uranium deposits. Rather, one must search for patterns within deposits that are of equal age (say within about 50 m. y.). When this is done, the Proterozoic is seen to comprise not one pattern but many, perhaps 40 or more.

Representative Proterozoic Uranium Deposits.

In order to determine how well the actual uranium deposits follow the patterns predicted by the Tectonospheric Earth Model, it is well to consider the locations and characteristics of representative examples of today's known Proterozoic uranium deposits in various parts of the world. This is done in the following sections.

Agnew Lake (Canada).

See Elliot Lake.

Alligator River Area (Australia).

The Alligators Rivers area, in Northern Territory, contains four main deposits. From north to south, these are: Nabarlek (q. v.) Jabiluka (q. v.), Ranger 1 (q. v.), and Koongarra (q. v.).

The geology of this area is dominated by two vestiges of seismotectonomagmatic-belt activity (i. e., the Nanambu and Nimbuwah complexes) surrounded by Proterozoic metamorphic rocks. The geological histories of these two complexes suggest two separate seismotectonomagmatic-belt episodes (Tatsch, 1972a: chapter 7). The granitoid core of the Nanambu complex is a vestige of Archean seismotectonomagmatic-belt activity, suggesting that the surrounding Proterozoic rocks are derivatives therefrom. The effects of the seismotectonomagmatic-belt episode of 1.8 b. y. ago are also seen.

The companion Nimbuwah complex, lying to the northeast of the Nanambu complex, comrpises an Archean core, part of which crops out as schist and quartzite mainly east of the East Alligator River. Some evidence suggests that these rocks are metamorphosed and mylonitized equivalents of the Proterozoic South Alligator group. In migmatized form, these constitute the outer zones of the Nimbuwah complex. Some evidence suggests that the Nimbuwah complex could have been an Archean core like that of the Nanambu complex. In any case, the effects of a seismotectonomagmatic-belt episode, at about 1.84 b. y. ago, correlate with the peak of migmatization and metamorphism.

Altered radioactive pink biotite granites appear to be anatectic and produced by migma development in the center of the Nimbuwah complex during migmatization. These are probably comagmatic with the Edith River volcanics, a series of acid volcanics that crop out in the South Alligator valley. They unconformally overlie folded South Alligator group sediments. These acid igneous rocks define the last Proterozoic episode of seismotectonomagmatic-belt activity.

Two series of basic igneous rocks intrude the Proterozoic rocks: (1) the Zamu Complex diorite, dolerite, and differentiates that were emplaced as numerous sills prior to migmatization and later altered, at places, to amphibolite; and (2) the Oenpelli Dolerite, emplaced, as a large undulating flat-lying discordant sheet, shortly after metamorphism. The Oenpelli Dolerite appears to have been emplaced before the migmatite complex had cooled.

See other details of this area under the headings, Rum Jungle - Alligator Rivers; Katherine Darwin. Other details are in the literature (See, for example, Ayres and Eadington, 1975).

Ambadungar (India).

Major fluorite deposits are associated with the uraniferous alkaline igneous rocks and carbonatites at Ambadungar, Gujarat. Similar associations occur at Sevattur, Tamil Nadu. The fluorite-bearing pegmatites of Bhilwara, Rajestan, contain many uranium minerals. The fluorite occurrences in Chandidongri, Madhya Pradesh, are matched by uranium occurrences in Dongargarh. Similarly, the fluorite mineralization in Chapoli, Rajestan, is matched by the high-temperature uranium mineralization in Khandela. These uranium-fluorite associations support the use of fluorite in the exploration for uranium. This would be particularly useful in those areas where uraniferous vestiges of seismotectonomagmatic-belt activity are suspected to be associated with fluorite vestiges (Tatsch, 1973a).

Anabar (U. S. S. R.).

The uraniferous mineralization in the basinal vestiges of Proterozoic episodes of seismotectonomagmatic-belt activity along the southern edge of the Anabar shield are reminiscent of similar mineralizations near other shield areas of the world (Tatsch, 1973a). The mineralization occurs mainly within the dike swarms that occupy the basins of the Keengeede and Kyuenelekeen rivers. Some of these dikes are almost 20 km long and 1/2 km thick. They dip at angles of 60° to 90° and comprise diabase, gabbro-diabase, and some granodiorite porphyry. The Proterozoic dikes cut the Archean schists and gneiss of the area.

Arnhem Land (Australia).

This deposit of the Northern Territory is part of one of the world's largest uraniferous areas (See, e. g., Langford) 1974). Arnhem Land is part of a large uraniferous province that extends eastward from Rum Jungle (q. v.) through the South Alligator district, to the Westmoreland and Mary Kathleen (q. v.) areas of Queensland. See also Kathleen-Darwin; Nabarlek.

Athapuscow Aulacogen (Canada).

The Athapuscow aulacogen contains several uraniferous deposits including Reliance (q. v.), Simpson Islands (q. v.), and Snowdrift (q. v.). The Athapuscow aulacogen is a vestige of seismotectonomagmatic-belt activity that occurred in the Canadian shield during the Archean and early Proterozoic (Tatsch, 1972a). Specifically, the Athapuscow aulacogen is the remnant of an incomplete seismotectonomagmatic belt that formed perpendicularly to the major early Proterozoic seismotectonomagmatic belt that ran in a northwesterly direction, from the Great Lakes to the Arctic Ocean. Besides these lakes, other vestiges of this early seismotectonomagmatic-belt activity include the lakes Great Slave, Athabasca, and Winnipeg (See,

Within the Athapuscow aulacogen, epigenetic uranium deposits occur within several areas of Proterozoic marine-to-continental volcanic-sedimentary sequences, including Reliance, Simpson Islands, and Snowdrift. Mineralization occurs in fautled and folded fluviatile arenites and rudites. Some of these have been partly metamorphosed. The sediments contain interstitial, disseminated uranium minerals. These have replaced pre-existing sulfides. The local basement comprises Archean and early Proterozoic metamorphic and plutonic rocks, ranging from 2.6 to 2.3 b. y. old. The Athapuscow aulacogen forms the boundary between the Slave and Churchill structural elements of the shield and is coextensive with a major system of high-angle faults that remain as vestiges of Archean seismotectonomagmatic-belt activity (Tatsch, 1972a). Unconformably lying above the Archean metamorphic and plutonic rocks are sequences of Proterozoic sediments and volcanics cut by intrusions associated with Aphebian and Helikian episodes of seismotectonomagmatic-belt activity (Tatsch, 1972a).

The vestigal remains of the Archean and early Proterozoic seismotectonomagmatic-belt activity now comprise 12 km of Proterozoic rocks in the southwestern part of the Athapuscow aulacogen where it intersects the previously-described northwesterly-trending vestiges of the Great Bear to Great Lakes seismotectonomagmatic belt. The Athapuscow shallows away from this line, being only about 2km deep near Fort Reliance.

The uraniferous fluviatile arenites of this area resemble the epigenetic sandstone-type deposits of other parts of the world. This is particularly noticeable with respect to the quartzite hosts that are found in the Black Hills of southwestern South Dakota and northeastern Wyoming, U. S. A. There is also a marked resemblance to some uraniferous deposits of Africa and the U. S. S. R. The Franceville (q. v.). deposits of Gabon are associated with a Proterozoic intracratonic basin, and the local redistributional features near some faults are reminiscent of the deposits of the Athapuscow aulocagen (See, e. g., Bournel and Pfiffelman, 1972).

Some observers feel that the Athapuscow aulacogen deposits may differ from similar deposits in other parts of the world, primarily in that the Canadian deposits might be unique examples of the sandstone type generated by metamorphic-fluid activity, rather than by the circulation of ground waters at shallow depths. In this respect, they could form a sub-class between the classic sandstone type of younger eras and the mobilized, synmetamorphic types such as the vein deposits at Eldorado (Saskatchewan) and Port Radium - Echo Bay (Northwest Territories).

Australia (Australasia).

The principal Proterozoic uraniferous deposits in Australia are: (1) Rum Jungle, 80 km south of Darwin; (2) South Alligator and East Alligator Rivers area in Northern Territory, 320 km east of Darwin; and (2) at Westmoreland, 700 km east of the Alligator rivers area.

Baker Lake (Canada).

This Northwest Territories uraniferous prospect is being tested to determine the origin, evolution, and present characteristics of the uranium content. Initial analyses suggest that this uranium is traceable to the same late Archean and Proterozoic seismotectonomagmatic-belt activity that accounts for the tectonics in this part of the Canadian shield (Tatsch, 1973a).

Bakouma (Central Africa Republic).

The uraniferous ores at Bakouma comprise phosphatic-filling karst in Precambrian dolomite (Sherman, 1972). They average about 0.10% U_3O_8.

Bancroft (Canada).

The deposits in this area are the only granite pegmatites that are worked extensively for uranium. Grading about 0.11% U_3O_8, the mineralization comprises uraninite and uranothorite plus some allanite, betafite, fergusonite, and zircon. The ore bodies are complex dikes and lenses that cut and replace highly metamorphosed Precambrian sedimentary rocks. The pegmatites form en echelon dikes along strike with vestiges of seismtoectonomagmatic-belt activity (Tatsch, 1973a).

Associated low-grade deposits, peripheral to the above, grade from 0.03% to 0.05% U_3O_8.

Bartalao (India).

The uraniferous Bartalao conglomerates occupy two andesitic flows in the Dongargarh area of Madhya Pradesh. The formations are part of a volcanic-sedimentary sequence of the Precambrian Dongargarh system lying about along the periphery of an eugeosyncline. They are underlain by rhyolites, rhyolitic conglomerates, and tuffaceous sandstones of the Bijli sequence. This rests on highly folded and metamorphosed gneisses and schists. Nearly is the Dongargarh granite considered to be intrusive into the Bijli sequence.

Uranium and thorium concentrations occur in these conglomerates at Bagadwani, Bartalao, Bunnara, Buranchhapur, Khampura, Kolarghat, Lachna, and Udrichhappar. Surficaal uraniferous concentrations of 0.02% to 0.03% U_3O_8 are interspersed by spots having values of 0.1% U_3O_8 (Udas and Mahadevan, 1974). The areas with the highest concentrations of thorium are associated with the presence of huttonite ($ThSiO_4$).

At Udrichhappar, the uranium concentration increases with depth, where there are coffinite and uraninite with pyrite. The underlying andesitic flow appears to have acted as a barrier during ground water action.

The Dongargarh granites and associated pegmatites, over which the andesitic flows and the Bartalao sandstones rest, contain disseminations of thorogummite, uranophane, and uranothorite. The rhyolitic tuffs and sandstones at the base of the Dongargarh system contain concentrations of 0.01% to 0.02% U_3O_8.

Beaufort West (Africa).

The Beaufort West uranium deposit lies within the Karoo supergroup of the Cape Province, South Africa (See, e. g., Von Backström, 1974). The mineralization lies within sedimentary rocks of the Lower Formation of the Beaufort group. The grade of the uranium content ranges from a trace to over 2% U_3O_8. The average is about 0.05%.

The surficial mineralization comprises bright green and yellow secondary uranium veins, primarily along joint planes. The mineralization lies within a sequence of glacial deposits, sandstones, mudstones, and shales, mostly continental, plus thick flows of lava, all intersected by and interlayered with dikes and sheets of dolerite. There are also workable seams of coal and fossil plants plus abundant remains of fish, amphibia, reptiles, and other vertebrates.

The urnaium mineralization is usually associated with deposits formed along erosion channels and washouts. The actual uranium occurs within thinly bedded lenses and joint fissures along the bedding planes. The association with organic remains is widespread. This includes fossil flora, fossilized bones, and carbonaceous material.

Specific mineralizations include a complex variery of secondary uraniferous minerals, including: (1) the bright-green metazeunerite; (2) the bright-yellow arsenuranylite; and (3) the pale-yellow uranophane. These appear to have been derived from the primary minerals, coffinite and uraninite. Associated minerals include pyrite, chalcopyrite, arsenopyrite, bornite, ilmenite, digenite, rutile, and martite. These minerals, primary and secondary, are associated with at least two vsstiges of late Archean and Proterozoic seismotectonomagmatic-belt activity (Tatsch, 1973a).

<u>Belo</u> <u>Horizonte</u> (Brazil).

The conglomerates of the Moeda formation south of the Sao Francisco craton are uraniferous. This formation lies sharply unconformably over highly metamorphosed rocks of the Nova Lima group. Overlying the Moeda is the Batatal formation, a graphitic siltstone, and then red hematitic iron ores of the Itabira group (See, e. g., Tatsch, 1975g).

The quartz-pebble conglomerates of the highly quartzose Moeda stretches linearly over a distance of about 50 km aligned with a vestige of seismotectonomagmatic-belt activity (Tatsch, 1975f). The rocks are greenish-gray to white and show the effects of metamorphism relatable to the intense thrust faulting associaate with the last episode of seismotectonomagmatic belt activity in that area.

The uraniferous minerals include brannerite, xenotime, and zircon. Balls of thucholite contain inclusions of uraninite with a relatively high thorium content, reminiscent of the thucholites in some of the Canadian pegmatites and the Witwatersrand conglomerates. Like the latter, some of the Belo Horizonte thucholites also carry significant amounts of gold (See, e. g., Tatsch, 1975f).

These rocks, like others of this area, were affected by an episode of seismotectonomagmatic-belt activity after about 1.93 b. y. ago. But the age of the conglomerates is older than this, perhaps dating back to episodes of seismotectonomagmatic-belt activity about 2.8 b. y., about 2.5 b. y. ago, and about 2.2 b. y. ago (Tatsch, 1973a). Other details of Belo Horizonte are contained in the literature (See, e. g., DeAndrade-Ramos and Fraenkel, 1974; Robertson, 1974).

In many respects the Belo Horizonte deposits resemble those of the Blind River - Elliot Lake district (q. v.).

Bhilwara (India).

See Ambadungar.

Bihar (India).

These uraniferous veins contain roughly 20,000 tons of ore averaging about 0.07% U_3O_8. See also Singhbhum.

Blind River - Elliot Lake (Canada).

These uraniferous deposits in quartz-pebble conglomerates contain uraninite and brannerite averaging 2 to 5 m thick and as much as 3 or 4 km across. They contain more than 5 million tons of ore. Minable grade ore averages 0.12% to 0.16% U_3O_8. The origin of the uranium is traceable to vestiges of late Archean and Proterozoic seismotectonomagmatic-belt activity in that area (Tatsch, 1973a).

Borborema (Brazil).

See Serido.

Botswana (Africa).

Yellow uranium oxides occur within the sandstones lying below the coal measures in North and South Botswana.

Brazil (South America).

Pyritic, quartz-pebble conglomerates with uranium and gold have been found in several areas along the periphery of the Ancient Sao Francisco craton (See, e. g., Tatsch, 1975f, 1975g). These areas include: (1) Cavalcante (q. v.) in Golias; (2) Belo Horizonte (q. v.) in Minas Gerais; (3) Pitangui (q. v.) west of Belo Horizonte; and (4) Jacobina (q. v.) in Bahia. Some uranium is found also in the potash-rich rhyolites of western Bahia, about 500 km southwest of

Jacobina (See, e. g., Sighinolfi and Sakai, 1974).

Brown (Australia)

See Rum Jungle.

Bunnara (India).

See Bartalao.

Burhanchhapur (India).

See Bartalao.

Cape Province (Africa).

See Beaufort West.

Carswell (Canada).

See Mokta.

Cavalcante (Brazil).

The uraniferous metaconglomerates in the northern part of the State of Golias lie at the northwestern boundary of the Sao Francisco craton. These conglomerates loverlie Archean micaschists and appear to be related to the same episode of seismotectonomagmatic-belt activity that produced the Minas series near Belo Horizonte (q. v.).

Cercado (Brazil).

This deposit resembles Agostinho (q. v.). The uraniferous mineralization occurs within a tectonic breccia of hydrothermally altered tinguaite of the Pocos de Caldas (q. v.) pipe. The average grade is about 0.18% U_3O_8 (Andrade-Ramos and Fraenkel, 1974). The uranium appears to have been derived from the metamict zircon of the area, probably by the action of hydrothermal metamorphism and associated action within that vestige of seismotectonomagmatic-belt activity (Tatsch, 1973a).

Charlebois Lake (Saskatchewan).

This deposit averages about 0.2% U_3O_8 in an area where biotite granite and biotite quartz monzonite intrude metamorphic rocks. The granite and quartz monzonite are cut by light-colored pegmaties of granitic composition. Uraninite is closely associated with the biotite.

Charmoli (India).

This Uttar Pradesh mineralization contains 0.02% to 0.18% U_3O_8 in association with minor sulfides in old copper workings and veins with sheared chlorite schists. The uraniferous solutions appear to have been derived from the granitic vestiges of seismotectonomagmatic-belt activity in that area (Tatsch, 1973a).

Chattanooga (U. S. A.).

The black shales of the eastern and central U. S. A. contain between 0.001% and 0.006% U_3O_8 (See, e. g., Bowie, 1970). The best known of these are the Chattanooga shales of Tennessee, Arkansas, and the adjoining states. The greatest concentration (0.006%) occurs within the 5-m upper member of east-central Tennessee, over an area of hundreds of km^2. The uranium is associated with vestiges of a Proterozoic episode of seismotectonomagmatic-belt activity in that area (Tatsch, 1973a).

Chaura-Tarandadhar (India).

This uraniferous deposit in the Himalayas resembles Wanaparthi (q. v.).

Chinjra (India).

This uraninite, near Kulu, occurs in veins. It occurs also, with pitchblende, as fracture fillings and coatings or replacements for quartz grains. The veins are generally parallel to the quartzite fold-fracture vestiges of seismotectonomagmatic-belt activity in that area (Tatsch, 1973a).

Cluff Lake (Saskatchewan).

See Mokta.

Coronation Hill (Australia).

This mine, part of the South Alligator River complex, contains pitchblende, some uranium secondary minerals, plus some minor pyrite and copper sulfides. The gangue is minor quartz veinlets. The host rock comprises tuffaceous rocks intimately associated with carbonaceous shale and siltstone. The mineralization comprises a sooty variety of pitchblende and some plitchblende veinlets. The ore body comprises a number of ore shoots (higher levels) and pipes (lower levels) extending about 60 m vertically and measuring about 25,000 tons of ore. The overlying igneous rocks are mainly rhyolite. Other details are included in the literature (See, e. g., Ayres and Eadington, 1975).

Cup Lake (Saskatchewan).

This area is part of the uraniferous vestiges of late Archean and Proterozoic seismotectonomagmatic-belt activity that trend northeasterly through Key Lake (q. v.) and Rabbit Lake (q. v.).

Dejour (Canada).

This new discovery in northern Saskatchewan is about 150 km southwest of Rabbit Lake (q. v.). The importance of this discovery is that it increases the potentialities of the entire area covering the southern edge of the Athabasca sandstone basin. It appears to be within part of the same vestige of seismotectonomagmatic-belt activity that contains Rabbit Lake (q. v.) (Tatsch, 1973a).

Dhabi (India).

Concentrations of coffinite and uraninite occur in the Precambrian syenitic rocks of Dhabi, Madhya Pradesh. These granitic gneisses dip steeply to the northwest on an east-west strike, reminiscent of uraniferous vestiges of seismotectonomagmatic-belt activity in other parts of the world (Tatsch, 1973a). They are traversed by tear faults and some oblique faults. The granite has been sheared along the faults, where subsequent hydrothermal activity has localized. These areas are now characterized by zeolitization of the feldspars and by deposition of arsenopyrite, pyrite, quartz, and uraninite. The uranium content is about 0.20% U_3O_8.

Some observers feel that late-stage shearing and hydrothermal activity are essential for the development of uranium concentrations in granitic rocks (See, e. g., Udas and Mahadevan, 1974). Examples of this in India include, besides Dhabi, the following: (1) the Chaura-Tarandhar area of the Himalayas; and (2) Dongargarh, Newania, and Wanaparthi in the Peninsular shield.

Dongargarh (India).

This uraniferous deposit in the Peninsular shield resembles Wanaparthi (q. v.). See also Ambadungar.

Dyson (Northern Territory).

See Rum Jungle.

Echo Bay (Northwestern Territories).

See Great Bear Lake.

Edmund (Western Australia).

See Mundong Well.

Eldorado (Canada).

See Great Bear Lake.

Elliot Lake (Canada).

This area of Ontario contains a small remnant of basal Proterozoic sediment beneath a much more widespread, younger Proterozoic sediment. The rocks of interest are those of the lower part of the Huronian supergroup. The outcrop area in which the uniniferous conglomerates are found is a 150km-long vestige of seismotectonomagmatic-belt activity stretching from Sault Ste. Marie to near Sudbury (Tatsch, 1973a). This vestige is all that outcrops from a longer vestige that once ran from Wisconsin to Labrador.

The economic conglomerates of the Huronian are at Elliot Lake and at Agnew Lake. The main ore minerals at Elliot Lake are brannerite and uraninite with secondary minerals such as gummite. Other heavy metals present include monzonite. The main ore mineral at Agnew Lake is uranothorite. Other heavy minerals there include monzonite, as at Elliot Lake.

The uranium appears to have been deposited syngenetically. The degree to which the ore consists of heavy mineral concentrate or of adsorption from solution is not known (See, e. g., Robertson, 1974). The ores total roughly 130,000 tons averaging less than 01.% U_3O_8 (Sherman, 1972).

El Sherana (Australia).

This mine, part of the South Alligator River complex, contains pitchblende, some secondary uranium minerals, and minor galena-clausthalite, pyrite, marcasite, Co-Ni arsenides, and Cu sulfides. The gangue is red and gray chert plus quartz veinlets. The host rock is primarily ferruginous siltstone and carbonaceous shale. The mineralization comprises lenticular pitchblende masses in a fault zone enveloped by abundant spherical nodules of pitchblende. The tabular ore body tapers with depth. The vertical extent is roughly 40 m. The ore comprises about 62,000 tons. The overlying igneous rocks are rhyolite. Other details are included in the literature (See, e. g., Ayres and Eadington, 1975).

<u>Ferghana</u> (U. S. S. R.).

See USSR Black Shales (Chapter 9), Ferghana (Chapter 6), and Tyuya Myuyun (Chapter 8).

<u>Finland</u> (Europe).

See Paukkajanvasra.

<u>Fort Dauphin</u> (Madagascar).

Uranium-bearing rocks are located along a 150-km by 35-km vestige of seismotectonomagmatic-belt activity stretching from Ft. Dauphin to the basin of the Mandrare River in southern Madagascar (Tatsch, 1973a). Uranothorianite is closely associated with masses and lenses of pyroxenite containing diopside and interstratified in Precambrian crystalline schists. Some of the lenses are several hundred m long. The uranothorianite occurs both in situ and as disseminations. In situ, it is associated with masses of anorthite, calcite, and phlogopite. The disseminations occur within the pyroxenite. Some alluvial and eluvial deposits, probably derived from the primary deposits, occur within the drainage area (Von Backström, 1974b). The drainage patterns are closely associated with vsstiges of seismotectonomagmatic-belt activity in that area (Tatsch, 1973a).

<u>Franceville</u> (Gabon).

Uraniferous vein deposits occur in the Franceville Precambrian sandstones and shales (1.745 b. y. old) that flank the Archean granite massifs. Pitchblende and coffinite form economic sedimentary deposits. These are further concentrated in structural and sedimentary traps. All deposits lie within 200 m of the present surface. The average grade is about 0.4% U_3O_8.

Fraserburg (Africa).

See Beaufort West.

Frome Lake (Australia).

These uraniferous deposits contain fairly large quantities of ore averaging better than 0.10% U_3O_8 (Sherman, 1971).

Gabon (Africa).

See Franceville.

Glenover (South Africa)

This carbonatite complex, in the Thabazimbi district, occupies a shallow, circular vestige of seismotectonomagmatic-belt activity, measuring about 13 km^2. The center of the plug is marked by 20-m high Breccia Hill. It comprises a high-grade apatite breccia. The two main types of carbonatites are beforsite and sövite. The carbonatite occurs as dikes and sills traversing the biotite-pyroxenite. The uranium grades only about 0.01% to 0.05% U_3O_8.

Great Bear Lake (Northwest Territories).

Pitchblende occurs at two locations near Echo Bay on the east shore of Great Bear Lake. One of the locations is 20 km north of Echo Bay; the other, about 175 km south. The pitchblende occurs with hematite in quartz veins (See, also, Tatsch, 1975g). Some of the deposits contain bismuth, chalcopyrite, dolomite, galena, and some Co-Fe-Ni minerals.

The uraniferous deposits in this area appear to have been affected by at least three Proterozoic episodes of seismtoectonomagmatic-belt activity (Tatsch, 1973a). These occurred about 2.5 b. y., 1.8 b. y., and 1.4 b. y. ago. Most observers feel that the uraniferous components of the deposits "originated" within the Echo Bay volcanics (1.8 b. y. old) and that they were transported by brines into fracture zones during post-Huronian (1.82 to 1.64 b. y. ago) regional faulting (See, e. g., Badham et al., 1972). Other details of the deposit are in the literature (See, e. g., Robinson and Ohmoto, 1973).

Gujarat (India).

See Ambadungar.

Gunnar (Saskatchewan).

This deposit is near the St. Mary's Channel of Lake Athabasca. It averages about 0.17% U_3O_8 (Von Backström, 1974b). The country rock is fine-grained paragneiss overlain unconformably by coarse-grained granitized gneisses and metasomatic granite. The uranium mineralization is within irregular, tabular-shaped bodies caused by albitization of the Gunnar granite. The orebody comprises brecciated and mineralized parts of the syenite mass. Ore minerals are pitchblende and uranophane. The uranium is extremely finely disseminated. Gangue minerals include calcite and hematite. Considered hydrothermal in origin, the uranophane appears primarily except near the surface. The enclosing breccia pipe appears spatially related to a number of vestiges of seismotectonomagmatic-belt activity (Tatsch, 1973a). Most of these are large faults, one or more of which probably provided conduits for the ore fluids.

Ilimaussaq (Greenland).

This mineralized lujauvritic nepheline syenite intrusion is

within an area of south Greenland that contains numerous vestiges of several episodes of seismotectonomagmatic-belt activity that occurred during the Proterozoic. These vestiges include a number of major intrusions, many dike swarms, and basaltic lavas. Ilimaussaq, emplaced roughly 1.0 b. y. ago, is the youngest of the major intrusions; it measures about 100 km^2. This intrusion has been enriched locally in thorium and uranium. The main minerals are eudialite, monzonite, and thorite. The uranium content ranges from 0.01% to 0.3%, mainly within low-grade, scattered mineralizations (Von Backström, 1974b). Other details are in Bohse et al. (1974).

Indus (Pakistan).

The Indus River, flowing from Tibet through Pakistan to the Arabian Sea, has deposited some contemporary sands that bear heavy minerals, including thorian uraninite (See, e. g., Cameron in Moghal, 1974). The Indus flood plains resemble the Tertiary Siwaliks associated with the deposits at Baghad Chur (q. v.).

Jabiluka 1 (Australia).

This deposit, in Northern Territory, is part of the Alligator Rivers area (q. v.). It lies roughly 50 km southwest of Nabarlek (q. v.) and roughly 500 km west of Jabiluka 2 (q. v.). The lode trends westsouthwest and dips south about 45°, in conformity with the vestiges of seismotectonomagmatic-belt activity in that area (See, e. g., Tatsch, 1973a). The ore zone extends along strike roughly 150 m at a depth of up to 100 m. The zone is about 30 m thick.

The primary ore is pitchblende in quartz-chlorite carbonaceous schist of the Koolpin formation equivalent. Minor constituents in the lode include pyrite and chalcopyrite (Tatsch, 1975g). Some gold is present locally. Some secondary uranium mineralozation is disseminated through a narrow zone in the Kombolgie formation sediments, apparently through hypogene enrichment. Reserves total about

3500 tons of U_3O_8 (See, e. g., Dodson et al., 1974; Anon., 1975a).

Jabiluka 2 (Australia).

This deposit, in Northern Territory, is part of the Alligator Rivers area (q. v.). It lies roughly 50 km southwest of Nabarlek (q. v.) and about 500 km east of Jabiluka 1 (q. v.). The mineralization underlies the Kombolgie formation sediments. The geology resembles that of Jabiluka 1; but the ore zone is larger, measuring roughly 50 m thick, at depths ranging from about 60 m to almost 200 m. Reserves are about 20,000 tons of U_3O_8, in ore that grades to 0.375% U_3O_8. Gold runs as high as 130 g/tn.

In the same area are several other anomalies, including Prospect 7J, where secondary uranium minerals crop out and gold runs as high as 23 g/tn.

Jachymov (Czechoslovakia).

This deposit comprises massive veins of pitchblende. The high-grade of this ore made it a chief source of radium. The mineralization appears to have been motivated by moderate to high temperatures and pressures that caused the lateral or upward flow of the uraniferous solutions along preferential paths of flow (Tatsch, 1972a). As is true with many other deposits of this type, the uranium is associated with bismuth, cobalt, and thorium.

Jacobina (Brazil).

The conglometates at Jacobina contain pyrite, gold, and some uranium. The enclosing rocks and the conglomerate that strike along the Serra de Jacobina over a distance of roughly 20 km are clean, well-washed, and recrystallized to a high degree (See, e. g., Robertson, 1974). The quartzites are overlain by hematite-bearing sediments, reminiscent of those of Itabira group deposits of Minas Gerais. The rocks appear to have been developed from debris shed from the Sao Francisco craton to the west (See, e. g., Robertson, 1974). The conglomerates appear to have been affected by the Trans-Amazon episode of seismotectonomagmatic-belt activity that occureed between 2.0 b. y. and 1.8 b. y. ago (Tatsch, 1972a).

Jaduguda (India).

These uraniferous veins contain over 15,000 tons of ore averaging roughly 0.07% U_3O_8.

Jhunjhnu (India).

This uraninite occurs within a shear zone of the Kolihan copper mine. The mineralization is in an amphibolite schist facies of argillaceous metasediments intruded by granite during a Proterozoic episode of seismotectonomagmatic-belt activity (Tatsch, 1973a).

Kalongwe (Zaire)

This deposit lies west of Shinkolobwe (q. v.) in southern Shaba (Katanga).

Karoo (Africa).

See Beaufort West.

Katherine-Darwin (Australia).

The Katherine-Darwin area comprises a large uraniferous province associated with vestiges of seismtoectonomagmatic-belt activity (Tatsch, 1973a). These vestiges extend eastward from Rum Jungle (q. v.) through the South Alligator district (q. v.) and on to the Westmoreland and Mary Kathleen (q. v.) areas in Queensland.

The uraniferous orebodies occur in the lower Proterozoic argillitic carbonaceous sediments that have mostly been metamorphosed to lower greenschist facies with extensive chloritic alterations (See, e. g., Langford, 1974). These were intruded by granites during an episode of seismotectonomagmatic-belt activity about 1.7 b. y. ago (Tatsch, 1973a).

Most of the Northern Territory uranium deposits are in lower Proterozoic rocks within 15 km of the middle and upper Proterozoic seismotectonomagmatic-belt outcrop. The larger of these deposits occur nearest the middle and upper Proterozoic vestiges of seismotectonomagmatic-belt activity (Tatsch, 1973a). This spatio-temporal correlation is usually expressed in the literature as an indication that the "first major regional control of uranium bodies occurred in lower Proterozoic rocks proximal to the regional unconformity". Under the seismotectonomagmatic-belt concept, this control is a vestige of seismotectonomagmatic-belt activity that occurred in that area during the late Archean (Tatsch, 1973a).

Examples of this Archean vestige and of early Proterozoic vestiges of seismotectonomagmatic-belt activity can be seen at Rum Jungle and in Arnhem Land, where the deposits appear to be closely grouped around the granite-gneiss complexes of these vestiges. In the South Alligator River area, the deposits lie along a fault zone associated with seismotectonomagmatic-belt activity.

Close scrutiny of these two controls shows that the uraniferous deposits are stikingly persistent in lying just below the unconformity. Almost all of the deposits are less than 100 m deep and are associated with some sort of dilatant shear or breccia zone associated with the vestiges of seismotectonomagmatic-belt activity.

<u>Kengeede</u> (U. S. S. R.).

See Anabar.

<u>Key Lake</u> (Canada).

This deposit, roughly 180 km north of Laronge, Saskatchewan, has a mineralized zone about 200 m by 30 m. The ore body, roughly 20 m thick, has a rich core surrounded by low-grade mineralization.

The uranium content ranges from 0.7% to 13.3% U_3O_8. The mineralization occurs in the Precambrian basement near contact with the younger Athabasca sandstones of the Wollaston basin. This vestige of seismotectonomagmatic-belt activity runs in a northeasterly direction to the area of Rabbit Lake (q. v.) and southwesterly to Cup Lake (Tatsch, 1973a).

<u>Khampura</u> (India).

See Bartalao.

<u>Khandela</u> (India).

See Ambadungar.

<u>Kolarghat</u> (India).

See Bartalao.

<u>Kombolgie</u> (Australia).

See Katherine-Darwin.

<u>Koolpin</u> (Australia).

This mine, part of the South Alligator River complex, contains pitchblende and secondary uranium minerals. The host rock is carbonaceous shale and ferruginous siltstone. The mineralization comprises sooty pitchblende and secondary uranium minerals in thin stringers occupying fractures. The ore body extends through a vertical distance of 15 m and measures about 2000 tons. The once-overlying middle Proterozoic igneous rocks have been removed by erosion. Other details are included in the literature (See, e. g., Ayres and Eadington, 1975).

<u>Koongarra</u> (Australia).

This deposit, in the Northern Territory, is part of the

Alligator Rivers area (q. v.). It lies roughly 20 km southsouthwest of Ranger 1 (q. v.), near the southern side of the Mt. Brockman massif, an outlier of Kombolgie formation sandstone. The mineralized zone conforms with vestiges of seismotectonomagmtic-belt activity within the area and approximately with the schistosity of the host rocks, i. e., parallel to a major reverse fault that dips about 60° to the southeast. The fault plane coincides with the southeast margin of the Mt. Brockman massif; this brings the lower Proterozoic host rock into contact with unmineralized Kombolgie formation sandstone (See, e. g., Dodson et al., 1974). The faulted contact is intensely brecciated and marks the footwall of the mineralized zone. The hanging wall is a carbonaceous horizon, roughly 70 m thick and almost parallel to the fault.

The orebodies comprise a series of enechelon zones of disseminated uraniferous minerals. The zones enclose cores of higher-grade ore. Host rock is lower Proterozoic quartz-chlorite-muscovite schist of the Koolpin formation equivalent. Accessory minerals include graphite and garnet. Pitchblende is the primary uraniferous ore mineral. Other associated minerals include chalcopyrige, galena, pyrite, and a trace of gold.

Krivoy Rog (U. S. S. R.).

This deposit, near Zheltye Vody, is in ferruginous rocks (See, e. g., Tatsch, 1975g). Uraninite occurs with sulfides of base metals. The mineralization occurs within the Saksagan granites (See, e. g., Gershoyg and Kaplun, 1970; Barnes and Ruzicka, 1972). Associated metals include copper, coablt, chromium, gold, lead, nickel, and silver.

Lachna (India).

See Bartalao.

Lake Baikal (U. S. S. R.).

See Slyudyanka.

Lesotho (Africa).

These uraniferous deposits are associated with the kimberlites that appear to be associated with vestiges of seismotectonomagmatic-belt activity in this part of southern Africa (Tatsch, 1973a).

Livingstonia (Malawi).

Uranophane occurs within sandstone of the Karoo system (q. v.), apparently above the coal horizon of the Middle Ecca series of this vestige of seismotectonomagmatic-belt activity.

Ljubljana (Yugoslavia).

These deposits resemble those at Zirovski Vrh (q. v.).

Madagascar (Africa).

The uranothorianite deposits of southern Madagascar have been fairly well worked out. The mineralization occurs in flat-lying masses of pyroxenite in Precambrian crystalline schists assocated with seismotectonomagmatic-belt activity in that area (Tatsch, 1973a). Some of the lenses are several hundred m long and 30 m thick. The average grade is less than 0.1% U_3O_8.

Mary Kathleen (Australia).

These uraniferous deposits of Queensland contain fairly large quantities of ore averaging better than 0.1% U_3O_8 (Sherman, 1971). See also Katherine-Darwin.

Medicine Bow (Wyoming).

The Precambrian rocks of the Medicine Bow Mountains and the Sierra Madre of southern Wyoming are being tested to determine the origin, evolution, and present characteristics of their uranium content (See, e. g., Anon., 1975b). Initial analyses suggest that this uranium is traceable to the same late Archean and Proterozocc seismotectonomagmatic-belt activity that accounts for the linearly contiguous uranium of this age in other parts of Wyoming and southern Canada (Tatsch, 1973a).

Merweville (Africa).

The pseudo-coal veins of the Merweville area of Cape Province contain some uranium. This mineraliza tin is probably associated with the same seismotectonomagmatic-belt episode that produced that of Beaufort West (q. v.).

Mikouloungou (Gabon).

The uranium ores at Mikouloungou comprise structurally controlled disseminations in Precambrian sandstone (Sherman, 1972). They average better than 0.1% U_3O_8.

Mokta (Saskatchewan).

Uranium is associated with copper, gold, nickel, platinum, selenium, and vanadium in three deposits of the Carwwell area. These are near the contact with the Archean basement and lie near Cluff Lake, on the southern side of the prominent circular Carswell structure, within the Athabasca sandstone basin. Possible interpretations for Carswell include cryptovolcanic, diapiric, and impact origins. Quasivolcanic Cluff breccis occupy dikes and veinlets in some areas. The richest of the three deposits averages 10% U_3O_8. The ore appears to be associated with vestiges of seismotectonomagmatic-belt activity, particularly a fault, in that area. The coffinite and uraninite are always associated with organic matter. Pitchblende occurs with shale in a shale-sandstone sequence that appears to have been overturned and overlain by basement (Von Backström, 1974b).

Montgomery Lake (Canada).

The Montgomery Lake - Padlei area, west of Hudson's Bay, represents part of the vestige of a large Proterozoic basin created by seismotectonomagmatic-belt activity during the late Archean or early Proterozoic (See, e. g., Tatsch, 1972a; 1973a). The rocks are part of the Hurwitz group and are normally red. In the Montgomery Lake - Padlei area, the Hurwitz rocks are unconformably underlain by a yellow-green to gray sequence of clastics lying on the Archean. The rocks, comprising a thickness of roughly 1 km, contain oligomictic quartz-pebble conglomerates with gold, pyrite, and uranium. The conglomerate beds range from less than a m to about 3 m in thickness.

The primary uranium mineral is uranothorite. The uranium is not as concentrated as is that at Elliot Lake (q. v.).

Mounana (Gabon).

The uranium ores at Mounana (Franceville) comprise structurally controlled disseminations in Precambrian sandstone (See, e. g., Sherman, 1972). The average grade is about 0.4% U_3O_8. See also Franceville.

Mudugh (Western Australia).

Carnotite is present interstitially and in cracks and cavities in calcrete (caliche), marls, clays, and bentonite in the near-surface zone. The uranium occurs from the surface to a depth of about 5-to-9 m. The primary source may be the granitic vestiges of the Archean episodes of seismotectonomagmatic-belt activity that outcrop in extensive zones of numerous basins, including the Yeelirrie basin (Dall'aglio et al., 1974; Tatsch, 1973a).

Mürtschenalp (Switzerland).

The Mürtschenalp uranium deposits are within the Canton of Glarus, located in the Helvetian nappe of the Verrucano province.

The sequence comprises (1) red conglomerate, (2) conglomeratic sandstone, and (3) mudstone. The sediments for this sequence appear to have been derived by arid erosion from vestiges of Variscan seismotectonomagmatic-belt activity (Tatsch, 1973a). Tectonism associated with this seismotectonomagmatic-belt activity produced intense jointing and the circulation of solutions that casued part of the sediments to be bleached to a gray color. One result of this activity was the formation of sericite and chlorite plus the development of carbonate and quartz veins. Accompanying this was a mesothermal or epithermal copper mineralization that resulted in the replacement of the pitchblende. The uranium appears to have been deposited under the influence of organic material.

Other uranium deposits in Switzerland show even stronger imprints from the tectonic and mobilization phases of seismotectonomagmatic-belt episodes, dating from the Proterozoic through the Paleozoic. See, e. g., St. Bernhard, in Chapter 8.

Mundong Well (Western Australia).

This fissure-type uraniferous mineralization is located near Edmund, Western Australia. The fact that this carnotite deposit occurs in caliche (calcrete) makes it of special interest (See, e. g., Langford, 1974). This deposit occurs within a shear zone, and the pitchblende fills a small fracture associated with seismotectonomagmatic-belt activity (Tatsch, 1973a). The host rock comprises quartzite metasediments in a Precambrian granite-gneiss complex. The deposit comprises fractures filled with botryoidal pitchblende, now significantly altered to gummite and malachite. It appears that this deposit was formed by surface waters carrying uranium through the fractured basement rocks of the vestiges of seismotectonomagmatic-belt activity that contained sulfides (Tatsch, 1973a). The sulfides supposedly provided the reducing environment that caused the precipitation of the uranium.

Nabarlek (Australia).

This deposit, in Northern Territory, is part of the Alligator Rivers area (q. v.). It comprises two high-grade lensoid lodes composed of ore richer than 10% U_3O_8. These lodes are surrounded by a zone of low-grade ore, roughly 230 m by 20 m. Massive pitchblende coated with secondary minerals extends from about 1 m to 70 m below the surface, where it lenses out above a thick flat dike of Oenpelli dolerite. Disseminated uraniferous minerals occur along strike for almost 300 m. The lodes strike northnorthwest parallel to the schistosity of the host rock and to other vestiges of seismotectonomagmatic-belt activity of that area (Tatsch, 1973a). The host rock comprises quartz-chlorite-muscovite schist of the Koolpin formation equivalent and dips roughly 30° to 45° eastward. The body of roughly 400,000 tons averages about 2.4% U_3O_8 (See, e. g., Dodson et al., 1974).

Namibia (Africa)

See South West Africa.

Newania (India).

This uranium deposit in the Peninsular shield resembles Wanaparthi (q. v.).

Nullagine (Australia).

The Proterozoic conglomerates near the town of Nullagine, Western Australia, contain uraniferous thucholites. They occur at the base of the Mount Bruce supergroup on the northern margin of the Hamersley Basin. This basin contains almost 4 km of basic pillow lavas and pyroclastics with a thin arkosic unit at the base. On the northern rim of the basin in the Nullagine and Marble Bar areas, the basal Fortescue is represented by a sequence containing well-developed conglomerates deposited in shallow depressions, structural basins, and other vestiges of seismotectonomagmatic-belt activity on the Archean land surface.

The conglomerate, where present, is overlain by up to 500 m to 600 m of acid-to-intermediate volcanics which elsewhere rest unconformably on the Archean. Overlying these volcanics are 1.5 km of shale and siltstone with occasional sandstone and conglomerate (Glen Herring), followed by the Green Hole formation.

Clasts, mainly vein quartz, vary from cm pebbles to 10-cm cobbles, all lying in a greenish-gray, arkosic matrix. The uranium ranges from a trace to 0.01% and is associated with thucholite pellets. Other heavy minerals are anatase, monzonite, pyrite, and zircon.

The basal rocks appear to have been deposited during seismotectonomagmatic-belt activity that occurred between 2.6 b. y. and 2.0 b. y. ago (Tatsch, 1972a). Overlying the drab-colored arkosic conglomerate sequence are younger rocks in which the iron forms red oxide (Tatsch, 1975g).

Oklo (Africa).

This 1.8 b. y. old deposit is unique in that it supports a natural reactor (See, e. g., Drozd et al., 1974; Lancelot et al., 1975). See also Franceville.

Onca do Pitangui (Brazil).

These low-grade deposits, about 100 km west of Belo Horizonte (q. v.), resemble the latter in most respects, but they are outside the Quadrilatero Ferrifero (Tatsch, 1975g). Apparently of the same Moeda formation, the deposits at Onca do Pitangui show larger thorium and titanium concentrations and are stratigraphically higher within the formation. The uranium appears to have been derived from the Sao Francisco craton.

Ondurakorume (South West Africa).

This carbonatite complex, about 55 km southwest of Otjiwarongo, projects as a prominent conical mountain roughly 300 m above the plain. The country rock comprises marble, schist, granite, and other formations of the Damara supergroup. The outcrop expression of this intrusive vestige of seismotectonomagmatic-belt activity is roughly circular and about 1 km in diameter. The composition of the intrusive is complex, including white sövite (oldest), radioactive greenish-black beforsite, brown beforsite, and some small radioactive gray-blue beforsite lenses and ring dikes. The greenish-black beforsite, which forms an irregular mass near the core and some minor outlying lenses and plugs, contains radioactive iron-rich veins and some apatite. The gray-blue beforsite is heavily impregnated with blue riebeckite, iron carbonate, strontium, and rare-earth carbonates. The uranium averages about 0.002% U_3O_8.

Padlei (Canada)

See Montgomery Lake.

Palette (Australia).

This mine, part of the South Alligator River complex, contains pitchblende, some secondary uranium mineralization, and minor galena-clausthalite, coloradoite, pyrite, marcasite, and gold. The gangue is largely apatite intruduced into the host sandstone. The mineralization comprises rich ore shoots along fault, plus spherical nodules of imperfectly developed pitchblende. The ore, which extends 30 m vertically, measures about 5000 tons. The overlying igneous rocks are rhyolites. Other details are contained in the literature (See, e. g.. Ayres and Eadington, 1975).

Paukkajanvaara (Finland).

This deposit occurs in northern Karelia within Precambrian quartzite-conglomerate formations reminiscent of those at Elliot

Lake (q. v.). But the mineralization occurs as veins and impregnations associated with diabase dikes. The ore zone parallels a dike and is best developed where the dike cuts a conglomerate bed. Pitchblende, the main uranium mineral, is associated with oxyhydrocarbons. The grade is about 0.2% U_3O_8. Other minerals include iron and vanadium.

Phalaborwa (South Africa).

This area comprises pyroxenite, syenite, olivine-diopside-phlogopite-pegmatoid, fennite, and carbonatite. Most of these are intrusive vestiges of seismotectonomagmatic-belt activity during the Archean and early Proterozoic (Tatsch, 1973a). The basement of Archean granite gneiss was intruded first by the pyroxenite, followed by the syenite, and lastly by a central core of transgressive carbonatite. This is surrounded by serpentine-magnetite-apatite rock, locally called "phoscorite", which has been enriched in magnetite and apatite (See, also, Tatsch, 1975g). Large concentrations of vermiculite occur in the northernmost pegmatoid body, and the carbonatite contains concentrations of copper sulfides. Some small concentrations of baddelayite and uranothorianite occur in the phoscorite and carbonatite. The uranium, which is mined as a by-product, averages about 0.004% (Von Backström, 1974b).

Pilanesberg (South Africa).

This low-grade mineralization, containing only about 0.005% to 0.025% U_3O_8, is within an intrusive complex rising sharply about 600 m above the Bushveld. Roughly circular with a diameter of about 27 km, this intrusive contains both intrusive and extrusive rocks, comprising mainly (1) alkaline syenites and trachytes, and (2) foyaites and phonolitic rocks. Some breccias and tuffs are associated with the extrusive rocks. Rare-earth oxides (up to 20%) occur as veins in tinguaite and as veins and blebs in the foyaites.

Pitangui (Brazil).

The undeformed, basal Proterozoic quartzites and arkoses of Pitangui are about 250 km northwest of Belo Horizonte (q. v.), on the southeastern edge of the Sao Francisco craton. They overlie the highly deformed schist and gneiss of the Nova Lima group. The Proterozoic rocks include quartz-pebble conglomerates carrying uranium and gold similar to those at Belo Horizonte (q. v.).

Most observers feel that, prior to about 2.2 b. y. ago, the Sao Francisco craton (1200 km by 600 km) was shedding clastic debris from its flank (See, e. g., Robertson, 1974). This is based on the assumption that the craton had uraniferous heavy minerals over much of its area and that these were concentrated by the major drainage patterns that reflect vestiges of seismotectonomagmatic-belt activity in that area.

Portugal (Europe).

Portugal's uranium deposits, which span the Precambrian to the Mesozoic, are discussed in Chapter 7.

Quadrilatero Ferrifero (Brazil).

See Belo Horizonte.

Rabbit Lake (Saskatchewan).

The Rabbit Lake uranium deposit lies within the Wollaston Lake fold belt in the southwestern part of the Canadian shield. This fold belt is a vestige of seismotectonomagmatic-belt activity within the Churchill structural province of the shield. During the lower Proterozoic, several northeasterly trending troughs formed on the Archean basement as part of that seismotectonomagmatic belt episode. The troughs were filled with sediments derived from the Archean uplands bordering its rims (See, e. g., Knipping, 1974; Tatsch, 1972a).

The Wollaston Lake fold belt, the most prominent of the lower Proterozoic sedimentary vestiges of that seismotectonomagmatic-belt activity, is a zone of dominantly supracrustal rocks. This vestige extends northeastward a distance of roughly 700 km from the southern edge of the Canadian shield to the border of Manitoba and the Northwest Territories. The width varies from 10 to 80 km.

Archean rocks, the source of the sediments within this vestige, are mostly granitic, as reflected in their arkosic character. The eugeosynclinal part of the seismotectonomagmatic-belt is represented in the southern segment by meta-arkose, hornblende-biotite rocks, acidic metavolcanics, and quartzites. The miogeosynclinal part of the seismotectonomagmatic-belt is represented in the Rabbit Lake area by meta-arkose, biotite paragneiss, and calc-silicate rocks with some marble.

The sedimentary assemblege was deformed and metamorphosed during the Hudsonian orogeny about 1.8 b. y. ago (See, e. g., Tatsch, 1972a: chapter 7). Vestiges of this seismotectonomagmatic-belt activity are seen today in the tight and isoclinal folds trending northeasterly. The associated metamorphism proceeded to the cordierite-amphibolite stage of the Abukuma facies series.

Many conformable and cross-cutting pegmatites are seen in the area. Foliated and unfoliated granites are there but not in the immediate vicinity of the Rabbit Lake deposit. The foliated granites appear to be remobilized Archean granites that intruded the lower Proterozoic sediments during the Hudsonian episode of seismotectonomagmatic-belt activity (roughly 1.8 b. y. ago) or during the earlier Kenoran episode of seismotectonomagmatic-belt activity (roughly 2.1 b. y. ago).

Unconformably overlying the lower Proterozoic is the middle and upper Proterozoic Athabasca formation. This, which is not metamorphosed, comprises quartz sandstone and conglomerate almost 2 km thick in certain places.

A few northwesterly trending diabase dikes intrude the above formations. The initial uranium mineralization appears to have occurred after the seismtotectonomagmatic-belt episode that produced these intrusions, probably about 1.075 b. y. ago. The time of this mineralization has been verified by other observers (See, e. g., Knipping, 1974).

The lower Proterozoic rocks south of Rabbit Lake are asymmetrically folded. The main structural feature is a northeasterly-trending steeply-dipping synclinal vestige of seismotectonomagmatic-belt activity. The axial region of this vestige was strongly sheared and brecciated. This is the ore zone at Rabbit Lake. Its general shape is that of an irregular pipe trending northeasterly and plunging gently to the northeast. A vestige of more-recent seismotectonomagmatic-belt activity, incluindg a northeasterly-trending cross-cutting reverse fault, limits the ore zone on the north end.

Alteration minerals include chlorite, dickite, kaolinite, and vermiculite. Also present are calcite, dolomite, hematite, and massive and euhedral quartz. Rutile, sphene, and other titanium minerals have been altered to anatase. Small amounts of magnesite and montmorillonite are also found. The alteration occurs only within the highly brecciated axial zone of the synclinal vestige and fractures associated therwwith. It is particularly intense within the upper 100 m and decreases rapidly downward to about 150 m, where it bottoms out into fresh unaltered rock. The interrelationships of the alteration minerals are very complex.

The uranium mineralszation appears to have begun after the alteration had been completed. At that time, the uraniferous solutions entered the porous rocks and deposited the minerals on the walls of the pores and fractures. The primary uranium minerals are colloform nnd massive pitchblende, sooty pitchblende, and some coffinite. Secondary minerals include boltwoodite, sklodowskite, and uranophane. The structural similarity of these is not too surprising.

Close scrutiny of Rabbit Lake and similar Proterozoic deposits in other parts of the world suggests that the paragenetic development within the orebody is always roughly the same and that it is a function of the nature of the associated vestige of seismotectonomagmatic-belt activity (See, e. g., Tatsch, 1973a). This paragenetic succession may be summarized:

a. Alteration of the host rock by the seismotectonomagmatic matic-belt activity.

b. Development of the porosity and permeability by the seismotectonomagmatic-belt activity.

c. Deposition of the pitchblende on the walls of the void spaces.

d. Deposition of massive quartz or calcite.

e. Deposition of euhedral quartz.

f. Deposition of euhedral dolomite.

g. Deposition of sooty pitchblende.

One or more of these steps may be missing from any particular deposit. If steps d and e (or either of them) are missing, the pitchblende might also be missing, because the coatings of massive quartz, calcite, or euhedral quartz serve to protect the pitchblende from leaching.

In using Rabbit Lake as the basis for finding similar deposits in other parts of the world, it is well to bear in mind that the evidence here suggests that the uranium mineralization is related to a major unconformity that remains as the most pronounced vestige of seismotectonomagmatic-belt activity in that part of the Canadian shield (See, e. g., Tatsch, 1973a). Other details are included in Drozd et al. (1974).

Rajasthan (India).

See Chapter 10.

Ranger (Australia).

This deposit, in Northern Territory, is part of the Alligator Rivers area (q. v.). It lies roughly 25 km south of Jabiluka (q. v.). The deposit comprises 6 radiometric anomalies extending roughly along a 7-km-long, north-trending arc about 1 km wide and in conformity with local vestiges of seismotectonomagmatic-belt activity (Tatsch, 1973a). Ore-grade uranium has been found at most of these anomalies.

The host rock appears to be part of the Koolpin formation equivalent. The mineralized zone comprises biotite-feldspar-quartz schist, overlain by a discontinuous band of dolomite and biotite-quartz schist. The schist in the vicinity of the mineralization is chloritized. Some carbonaceous schist is present locally. Some pegmatite and dolerite intrusions are heavily altered within the mineralized area.

The primary uranium occurs within pitchblende-rich chloritic veins filling cracks and fissures. In the upper parts, the uraniferous veinlets tend to concentrate around fragments of brecciated chlorite schist. Some pitchblende occurs as fine grains disseminated through veins and within mineralized zones of chloritic host rocks. Some copper, gold, and lead are sporadically didstibuted through the ore. The reserves total over 100,000 tons U_3O_8. See also Katherine-Darwin.

Reliance (Canada).

The Reliance mineralization occurs within the Kluziai formation, east of Meridian Lake. Reminiscent of the Simpson Islands (q. v.) mineralization, the Reliance deposits comprise 3 zones that occur within dark-red, hematitic arenites carrying some yellow uranium minerals on the fractures and joints (See, also, Tatsch, 1975g). Early tests show that the U_3O_8 content ranges from roughly 2.6% to 6.7% (See, e. g., Morton, 1974).

Interstitial lamellar hematite is accompanied by spherules and shells of uraninite and some brannerite. Some of these have been replaced by secondary coffinite. The uranium oxides have replaced primary pyrite and chalcopyrite. Some uraninite cubes are rimmed by coffinite. There are also calcite, cobaltite, and safflonite.

The Reliance deposits appear to have been epigenetically emplaced from solution roughly 1.55 b. y. ago to 1.80 b. y. ago. This age compares with 1.62 b. y. for the earliest stages of the Port Radium (q. v.) mineralization on the eastern shore of the Great Bear Lake (See, e. g., Thorpe, 1971).

According to the Tectonospheric Earth Model, the Reliance mineralization, along with other Proterozoic unanium deposits of similar ages, are vestiges of seismotectonomagmatic-belt activity that occurred during the late Archean and early Proterozoic. In that area of Canada, these vestiges are identifiable with the "lake line" that now strikes northwesterly and comprises the Great Bear, Great Slave, Athabasca, Winnipeg, and the Great Lakes. Part of the uranium, under this concept, is derived from detrital uranium from the Archean craton. Other contributions came, under this concept, from (1) exhalative uranium supplied to the seismotectonomagmatic belt during one of its volcanic episodes; (2) uranium leached from arkosic sands and volcanics of the Proterozoic seismotectonomagmatic-belt vestiges, during diagenesis and burial metamorphism; and (3) uranium that escaped from vestiges of the Churchill seismotectonomagmatic-belt activity, during the late (retrograde) stages of metamorphism and extreme cataclasis of the Hudsonian episode of seismotectonomagmatic-belt activity (Tatsch, 1973a).

One possible scenario involving specific vestiges of Archean and early Proterozoic episodes of seismotectonomagmatic-belt activity may be considered:

a. Roughly 2.00 b. y. ago. Erosion from the Archean basement deposited ilmenite and magnetite sands into the vestiges of the earlier seismotectonomagmatic-belt episodes (See also Tatsch, 1975g).

b. Roughly 1.87 b. y. ago. The seismotectonomagmatic-belt episode at this time included extensive submarine eruptions. These caused metallogenic fluids to pervade the sedimentary pile. Besides uranium, the fluids probably contained As, Ca, Co, Cu, Fe, and S. The oxides magnetite and ilmenite were converted by H_2S to pyrite and anatase (Tatsch, 1975g). This process was aided and abetted by some biologic activity or by the breakdown of sulfur-bearing organics.

c. Roughly 1.60 to 1.50 b. y. ago. The long sedimentary history within the vestige of the seismotectonomagmatic belt resulted in the burial of the lowermost Proterozoic sediments and associated volcanics under about 10 km of sediments and ingeous rocks. The present deposits resulted from post-seismotectonomagmatic-belt metamorphism.

Other details of the Reliance deposit are discussed under Athapuscow Aulacogen (q. v) and in the literature (See, e. g., Morton, 1974).

Rhodesia (Africa).

Uranium occurs within post-Karoo faults, apparently derived from Karoo beds that are associated with Archean and early Proterozoic seismotectonomagmatic-belt activity in that area (Tatsch, 1973a).

Richardson (Ontario).

An interesting feature of this mine is the association of brannerite with native gold (See, e. g., Steacy et al., 1974; Tatsch, 1975f).

Rockhole (Australia).

This mine, part of the South Alligator River complex, contains pitchblende plus minor clausthalite, eskebornite, pyrite , marcasite, chalcopyrite, and some rare secondary uranium minerals. The gangue is siderite. The host rock is primarily carbonaceous shale and chert (lower levels) and sandstone near the unconformity (upper levels). The mineralization comprises ore shoots in ribbon form dipping westward along fault zones. The orebody, which extends about 60 m vertically, comprises over 13,000 tons. The overlying rhyolite is in fault contact with the host rocks. Other details are in the literature (See, e. g., Ayres and Eadington, 1975).

Rössing (South West Africa).

This deposit is similar to Conway (q. v.) but it is more concentrated and less extensive. It occurs in alaskite rock. Averaging about 0.04% U_3O_8, this deposit comprises uraninite in fine grains and occurring as inclusions in quartz, feldspar, and biotite. The host rock has been dated at about 500 m. y. (Armstrong, 1974). The mine is scheduled to go on stream this year (1976).

Rum Jungle (Australia).

This area of Northern Territory is part of the Rum Jungle - Alligator Rivers complex (q. v.). See also Katherine-Darwin.

The Rum Jungle - Alligator Rivers complex comprises three uraniferous areas within Northern Territory: Alligator Rivers (q. v.), Rum Jungle, and South Alligator River Valley (q. v.). This is one of the most important uranium fields of the world. As is true of other large Proterozoic uranium deposits of the world, there is a distinct association between the pattern of the mineralization and the vestiges of Archean and early Proterozoic seismotectonomagmatic-belt activity in that area. This association is particularly noticeable with respect to the lower Proterozoic sedimentation that is a product of these vestiges.

With very few exceptions, all known uranium deposits of this complex occur in the lower Proterozoic sediments of the Pine Creek geosyncline, which is a direct vestige of an identifiable seismotectonomagmatic belt of that area (Tatsch, 1973a). Most observers feel that the most important clue to the origin of the uranium in this area is the recognition of a stratigraphic control on the distribution of the uranium (See, e. g., Dodson et al., 1974).

Present evidence shows that metasediments originally considered to be Archean are the metamorphosed equivalents of lower Proterozoic sediments (e. g., South Alligator group) exposed in the South Alligator River Valley. The concentration of the uranium suggests syngenetic enrichment during sedimentation. The exact source of the uranium is not known, but it appears that the provenance contained sufficient uranium to provide it during sedimentation. Some of the Archean rocks at Rum Jungle have uranium contents up to 0.0028% U_3O_8 (See, e. g., Heier and Rhodes, 1966).

Most orebodies occupy breccias, faults, and shears. Some, like the deposit at Rum Jungle Creek South, occupies a fold. Thus, it appears that the mineralizing solutions have been squeezed out from a sedimentary pile into centers of low stress during an episode of seismotectonomagmatic-belt activity.(Closer scrutiny shows also that the mineralization might be related to the seismotectonomagmatic-belt episode that was responsible for the rejuvenation of Archean igneous masses and for the intrusion of other igneous masses about 1,8 b. y. ago (See, e. g., Tatsch, 1973a). This latter episode is correlatable with seismotectonomagmatic-belt activity in other Precambrian shields (Tatsch, 1972a: chapter 7).

Extensive alteration zones around the deposits of the Alligator Rivers area suggests a mesothermal mineralization temperature range. This area shows vestiges of four episodes of seismotectonomagmatic-belt activity: 1.9 b. y., 1.7 b. y., 0.9 b. y., and 0.5 b. y. ago. These ages support the concept of syngenetic deposition of uranium in the lower Proterozoic sediments prior to 1.9 b. y. ago and uranium concentration at 1.7 b. y. ago, during the remobilization of pre-existing uranium accumulations by the seismotectonomagmatic-belt episode at 1.7 b. y. ago. This was followed by lead loss during the seismotectonomagmatic-belt episode at 0.9 b. y. ago and at 0.5 b. y. ago. Not all evidence supports full-fledged seismotectonomagmatic-belt episodes at 0.9 b. y. and 0.5 b. y. ago (See, e. g., Dodson et al., 1974).

The uranium deposits of the South Alligator River Valley are spatially closely associated with the Edith River volcanics. But most evidence suggests that these volcanics were not responsible for the deposits. More probably, the mineralizing solutions were fluids isolated by lithification of the lower Proterozoic sediments and mobilized during the igneous and tectonic phases of the seismotectonomagmatic-belt activity. Some supergene enrichment has occurred locally, probably by solution and accretion as the surface has been lowered by erosion; e. g., Palette in the South Alligator River Valley.

In summary, the following sequence of events appears to have been responsible for the formation of the uranium deposits of the

Rum Jungle - Alligator Rivers complex:

a. Syngenetic enrichment of uranium in carbonaceous sediments during the lower Proterozoic.

b. Concentration of ore-grade uranium minerals from mineralized solutions mobilized during a major episode of seismotectonomagmatic-belt activity.

c. Relocation of uranium by circulating solutions through leaching and redeposition nearby.

Other details of this deposit are contained in the literature (See, e. g., Ayres and Eadington, 1975).

<u>Russia</u> (Eurasia).

See USSR.

<u>Saddle Ridge</u> (Australia).

This mine, part of the South Alligator River complex, contains only secondary uranium minerals plus some pyrite. The host rock is early Proterozoic siltstone and middle Proterozoic tuffaceous sandstone and volcanics. The mineralization fills joints and fractures through a vertical extension of 25 m and comprises roughly 30,000 tons of ore. The nearest igneous rocks are the middle Proterozoic tuffaceous rocks. Other details of this deposit are in the literature (See, e. g., Ayres and Eadington, 1975).

<u>Sakami Lake</u> (Canada).

A thin sequence of rocks on the west side of Sakami Lake, in northern Quebec, contains a uraniferous quartz-pebble conglomerate (See, e. g., Robertson, 1974). The rocks vary from yellow-gray, coarse quartzite to white, fine-grained quartzite. Some beds have as much as 10% pyrite.

The rocks resemble the basal part of the Huronian supergroup. They lie unconformably atop folded greywacke and greenstone of Archean age. Above this is a redbed of conglomerate, arkose, and siltstone. The uraninite contains about 10% ThO_2. The uranium content is relatively high locally.

The basement rocks appear to have been affected by the Kenoran episode of seismotectonomagmatic-belt activity. The uraniferous conglomerates are intensely folded and also appear to have been involved in the Kenoran activity. They resemble those at Elliot Lake (q. v.).

Sandfly Lake (Saskatchewan).

This area is part of the uraniferous vestige of late Archean and Proterozoic seimsotectonomagmatic-belt activity that trends northeasterly through Key Lake (q. v.) and Rabbit Lake (q. v.).

Scinto 6 (Australia).

This mine, part of the South Alligator River complex, contains only secondary uranium minerals. The host is rhyolite. The mineralization comprises uranium ochres filling joints and fractures through a ventical extent of 35 m. The overlying igneous rocks are rhyolite. Other details are in the literature (See, e. g., Ayres and Eadington, 1975).

Serido (Brazil).

This area comprises a complex of geosynclinal vestiges of episodes of seismotectonomagmatic-belt activity in northeastern Brazil. It includes the Borborema uraniferous scheelite province. The basement discordantly overlies the Archean. The associated vestiges of seismotectonomagmatic-belt activity include at least 4 types of granitic rocks plus some basic and ultrabasic Proterozoic rocks as dikes and sills. The uranium mineralization is mainly (1) disseminated in granite and pegmatoid granite and (2) contained in veins localized in fracture zones and without lithological control. The primary uranium minerals within the disseminations are uraninite and uranothorianite; the secondary minerals include beta-uranophane, meta-autunite, meta-torbernite, and francolite. The primary uranium minerals in the veins are pitchblende and uraninite; the secondary minerals include uranophane, meta-autunite, kasolite. The average grades are 0.04% U_3O_8 for the disseminations; 0.2%, for the veins. Other details are contained in the literature (See, e. g., DeAndrade-Ramos and Fraenkel, 1974).

Sevattur (India).

See Ambadungar.

Shinkolobwe (Zaire).

This deposit is now exhausted, but it was part of the complex of uranium deposits that lie along the Zambesi-Damaran vestige of seismotectonomagmatic-belt activity (Tatsch, 1973a). This same belt extends southwesterly through Angola and into South West Africa.

Sierra Madre (Wyoming).

See Medicine Bow.

Simpson Islands (Canada).

The Simpson Islands are located about 140 km southeast of Yellowknife. Roughly 200 occurrences of uranium mineralization have been found in the Hornby Channel formation strata which unconformably overlie Archean biotite granites and upper amphibolite facies paragneisses with pegmatites (See, e. g., Morton, 1974).

Almost all (85%) of the uranium mineralization occurs within the lower half km of a 1.5-km-thick succession of mature, thick-bedded to massive, poorly-sorted, coarse orthoquartzitic and subarkosic sandstones, grits, and conglomerates. The mineralized members are noticeably deficient in fresh detrital magnetite and ilmenite, but contain numerous zircons (See, also, Tatsch, 1975g).

Locally, the beds are cut by diatreme breccias, diabase, syenite, and bostonite dikes. Silicification and stringers of quartz veins are found near faults striking roughly northeast-southwest.

The uraniferous zones are expressed at the surface by red hematite enrichment or by outcrops bearing yellow and green secondary uranium minerals. Beneath the surface, the deposits are epigenetic and discordant. The U_3O_8 content ranges up to 1.5% within dark gray, sulfide rich arenites and rudites.

Close scrutiny of the mineralized strata shows that the interstitial euhedral pyrite and chalcopyrite have been replaced by fine-grained anhedral uraninite. This has been replaced by anhedral coffinite-bearing euhedral galena. Some primary sphalerite and cobaltite occur with the sulfides. The uraninite accompanies balded hematite and euhedral-to-anhedral anatase (Tatsch, 1975g).

Other details of this deposit are discussed under Athopuscow Aulacogen (q. v.) and in the literature (See, e. g., Morton, 1974).

Singhbhum (India).

Uranium occurs in the well-known Singhbhum thrust belt in Bihar in close association with copper mineralization (See, e. g., Tatsch, 1975a). The mineralization extends into the Khetri-Dariba (q. v.) area of Rajasthan to the west. Besides copper and uranium, this uraniferous vestige of seismotectonomagmatic-belt activity contains also the Iron-ore series, Gangpur series, Sakoli series, and Aravalli and Delhi systems. Numerous Cu-Pb-U-Zn mineralizations are seen along this belt.

The mineralization areas are primarily base-metal and belong to the hypothermal-mesothermal-epithermal type. The associated vestiges of seismotectonomagmatic-belt activity are related primarily to a seismotectonomagmatic-belt episode involving early to middle Precambrian fold movements (Tatsch, 1972a). An analysis of the metallotectonic controls in that area has been made by Udas and Mahadevan (1974). See also Khetri-Dariba.

Skull (Australia)

This mine, part of the South Alligator River complex, contains pitchblende, secondary uranium minerals, and minor gold and copper mineralizations. The host rock is sandstone without any introduced gangue. The mineralization comprises pitchblende veinlets, nodules, and sooty pitchblende. The ore body measures roughly 540 tons. The overlying igneous rocks are rhyolite. Other details are included in the literature (See, e. g., Ayres and Eadington, 1975).

Sleisbeck (Australia).

This deposit, just over 30 km along strike from the South Alligator River deposits (q. v.) lies in carbonaceous shale and appears to be part of the same late Archean and Proterozoic vestiges of seismotectonomagmatic-belt activity that produced other Australian uranium deposits of that age (Tatsch, 1973a). Other details are included in the literature (See, e. g.. Ayres and Eadington, 1975).

Slyduyanka (U. S. S. R.).

This uraniferous deposit, in the Lake Baikal area, occurs in betafite. The deposit is primarily phlogopite mica with the betafite occurring only in pegmatite veins. The betafite is mined for the nionium and tantalum content, with uranium as a byproduct. Similar deposits occur at Betafo, Madagascar.

Snowdrift (Canada).

The uraniferous sediments of the Snowdrift area occur around Toopan Lake, roughly 20 km eastsoutheast of Snowdrift. Three mineralized zones, plus some showings, occur along a 3 km strike. Discordantly emplaced, the uranium deposits lie within a 180-m thick sequence of massive to thick-bedded, well-sorted, medium-fine grained, non-marine orthoquartzite of the Kluziai formation. The host strata contain miscellaneous green shale and shale breccias.

Reminiscent of the Simpson Islands (q. v.) mineralization, the Snowdrift deposit comprises three types of low-grade deposits: (1) a reduced sulfidic comprising graphite, uranium-titanium minerals, pyrite, chalcopyrite, cobaltite, and barite; (2) an oxidized type comprising uranium minerals with abundant hematite; and (3) a supergene-leached type with pale buff arenites, controlled by joint sets, and carrying yellow uranium minerals and goethite. Type (2) forms coronas around type (1). Type (3) replaces types (2) and (3). The U_3O_8 contents are roughly 0.5% for type (1), 0.1% for type (2), and 0.05% for type (3).

The chalcopyrite and pyrite are partly replaced by brannerite and a mineral resembling uranium anatase. Other details are discussed under Athapuscow Aulacogen (q. v.) and in the literature (See, e. g., Morton, 1974).

South Alligator River Valley (Australia).

This area of Northern Territory is part of the Rum Jungle - Alligator Rivers complex (q. v.). See also Katherine-Darwin, Coronation Hill, Palette, Rockhole, El Sherana, Saddle Ridge, Scinto 6, Skull, and Koolpin.

South West Africa (Africa).

See Rössing.

Swambo (Zaire).

This deposit lies west of Shinkolobwe (q. v.) in southern Shaba (Katanga) and appears to be related to the same vestiges of seismotectonomagmatic-belt activity that produced the uranium at Shinkolobwe (Tatsch, 1973a).

Tamil Nadu (India).

See Ambadungar.

Tarandadhar (India).

This uranium deposit in the Himalayas resembles Wanaparthi (q. v.).

Toopon Lake (Canada).

See Snowdrift.

Turiy Peninsula (U. S. S. R.).

This uraniferous deposit is associated with thorium in the alkalic rocks in the Murmansk area of the Kola peninsula (See, e. g., Bulakh et al., 1974; Tatsch, 1973a).

Tweerivier (South Africa).

This carbonatite complex, in the Brits district, measures about 6 km by 2 km and has a pear shape. It is entirely surrounded by red Bushveld granite. There are three volcanic vents in the area. The northern part of the complex comprises a heterogenous assemblage of dolomitic carbonatite into which an outer group of dikes was intruded. The southern part of the complex comprises altogether different rock types, i. e., gabbro, anorthosite gabbro, white crystalline limestone, and a radioactive silicified ferruginous rock. The uranium averages only about 0.002 to 0.006% U_3O_8 (Von Bockström, 1974b).

USSR Black Shales (U. S. S. R.).

These uraniferous black shales are probably associated with the same vestiges of seismotectonomagmatic-belt activity that produced similar deposits in other parts of the world; e. g., Sweden, and the central and eastern states of the U. S. A. (including Tennessee, Oklahoma, and Michigan). Within the U. S. S. R., these uraniferous vestiges of Proterozoic seismotectonomagmatic-belt activity occur mainly in the region just south of the Gulf of Finland. Other black shales occur in the Kara Tau Mountains of Kazakhstan, Ferghana in the Tien Shan, and the Caspian Sea area.

Wanaparthi (India).

Uraniferous concentrations occur within the Precambrian granites of this area of the Peninsular shield. The uraniferous minerals, coffinite and uraninite, occur within vugs along fractures, together with arsenopyrite, pyrite, and fluorite. Some secondary minerals occur also along joints and other fractures, probably as oxidation products from the primary uranium oxides in the rocks. See also Dhabi, which somewhat resembles this.

Westmoreland (Queensland).

See Mary Kathleen.

White (Northern Teritory).

See Rum Jungle.

Witwatersrand (Africa).

The Witwatersrand sequence of clastic sediments and interlayered volcanics contains gold and uranium. These metals occur in the Witwatersrand "reefs", which are actually conglomerates and thin grit bands, at various levels of the 7.5-km-thick section.

The enclosing arkose and quartzite together with interbedded silt, are yellowish-gray to white. The dip is low with almost no metamorphism. The basement is complexly distorted and metamorphosed. The Witwatersrand conglomerates are overliin by rocks of the Transvaal supergroup, the clastics of which are red and contain much of South Africa's iron ore reserves (See, also, Tatsch, 1975g).

The outcropping conglomerates of the Central Rand resemble those of Elliot Lake (q. v.). Like many others of the world, they could easily owe their existence to the same Archean episode of seismotectonomagmatic-belt activity (See, e. g., Tatsch, 1973a). In the case of Witwatersrand, the conglomerates appear to occupy part of the vestige of an early seismotectonomagmatic belt.

The main ore minerals are native gold and uraninite, both associated with thucholite. Other uraniferous minerals are uranothorite and zircon. Monzanite abounds only in the Dominion Reef (and possibly in the Central Rand).

The lavas of the Dominion Reef system are associated with an episode of seismotectonomagmatic-belt activity about 2.8 b. y. ago; the Ventersdorp lavas, with one about 2.3 b. y. ago; and the Transvaal group, with one between 2.3 b. y. and 2.0 b. y. ago.

It is of interest that the uraniferous deposits of Witwatersrand, like the gold deposits there, are detrital (See, e. g., Pretorius, 1974). These Proteroziic deposits have been redistributed by subsequent seismotectonomagmatic-belt activity (Tatsch, 1973a). The gold appears to have been derived from the greenstone vestiges of a seismotectonomagmatic-belt episode about 3.3 b. y. ago. The uranium appears to have been derived from the granitic intrusions associated with a seismotectonomagmatic-belt episode about 3.1 b. y. ago.

In this connection, it is of interest that the African shield appears to have stabilized earlier than did the other shields of the world. Also, greater volumes of mafic and ultramafic volcanics were associated with the 3.3 b. y. ago episode than with the 3.1 b. y.

ago episode of seismotectonomagmatic-belt activity (Tatsch, 1973a). In this respect, the African shield differs from the Canadian shield (Tatsch, 1972a), in accordance with the predictions of the Tectonospheric Earth Model.

Conclusions Regarding the Proterozoic Uranium Deposits.

Close scrutiny of the known Proterozoic uranium deposits reveals that almost all of them appear to be associated with those areas of the Earth's surface where vestiges of Proterozoic seismotectonomagmatic-belt activity remain today. These areas are found primarily within the platforms and their folded peripheries, including those of western Siberia, southern and western Africa, the United States, Canada, Brazil, India, Scandinavia, and Australia.

Chapter 10

ARCHEAN SEISMOTECTONOMAGMATIC BELTS AND THE ASSOCIATED URANIUM DEPOSITS

In considering the Archean uranium deposits, it is well to recall the basic enigmas that exist within our knowledge of the Archean. These include: (1) the remarkable similarities between the South African, Canadian, and Australian Archean shields (See, e. g., Oversby, 1975); (2) the differences between the Archean shields and younger rocks; (3) the evidence that the geological processes operating during the Archean were not exactly the same as they are today (See, e. g., Anhaeusser et al., 1969); (4) the large volumes of acidic and migmatitic rocks that were produced at a time when the Earth is usually assumed to have had only a thin, primitive crust; and (5) the fact that the source of material for the Archean rocks, particularly the more potassic ones, is not obvious when considered in connection with traditional Earth models. These and other Archean enigmas contributed, some years ago, to the author's realization that the early history of the Earth can be understood only in terms of a deep-seated, global driving mechanism that has been operating during the 4.6 b. y. that the Earth is believed to have been in existence (See, e. g., Tatsch, 1963a).

The uranium deposits associated with the Archean seismotectonomagmatic belts comprise, according to the Tectonospheric Earth Model concept, part of the Earth's "original", early-Archean uranium that was reworked by subsequent seismotectonomagmatic-belt activity during the Archean but not during the Proterozoic and Phanerozoic. It is not too surprising that the Earth's Archean uranium deposits are found primarily within the shields of Africa, Asia, Australia, and the Americas. These are the main areas where vestiges of Archean seismotectonomagmatic-belt activity still remain today. These vestiges occur mainly as long, rectilinear belts crisscrossing the

Archean cratons (See, e. g., Oversby, 1975).

In considering the distribution patterns of the Earth's Archean uranium deposits, it is well to bear in mind that the late-Archean shields were not necessarily the same as they are today, nor the same as they were during the early Archean. Thus, the Canadian shield could have been separated into two or more parts during the late Archean and early Proterozoic. For example, an archeoplate comprising the Churchill, Superior, and Southern provinces was probably separate from an archeoplate comprising the Bear and Slave provinces (See, e. g., Symons, 1975). Likewise, the archeoplates in other parts of the world must be analyzed in a similar manner before analyzing the origin and evolution of today's uranium deposits.

Eurasia provides an excellent example. Today, Eurasia comprises a multiplicity of cratonic units of various ages (See, e. g., Tatsch, 1972a: chapter 7). These cratons include (1) the north European plate, (2) the south European plate, (3) the Siberian platform, (4) the Jano-Kolymian plate, (5) the Kazakhstan plate (6) the north Chinese plate, and (7) the south Chinese plate. Separating these cratonic units from each other, and traversing at least some of them intracratonically, are fossil orogens, sutures, and other vestiges of erstwhile seismotectonomagmatic-belt activity dating back to the Archean. These 7 plates could easily have been parts of separate continental entities at least once during the Archean and perhaps many times during the Proterozoic and Phanerozonic (Tatsch, 1972a: chapter 7).

Representative Archean Uranium Deposits.

In order to determine how well today's actual uranium deposits follow the patterns predicted by the Tectonospheric Earth Model, it is well to consider the locations and salient characteristics of representative examples of known Archean uranium deposits in various parts of the world. This is done in the following sections.

Alligator Rivers (Australia).

The Alligator Rivers district comprises several deposits, including Jim Jim (q. v.), Rum Jungle (q. v.), Katherine-Darwin (q. v.), Nabarlek (q. v.), and Ranger (q. v.). Most of these average better than 0.1% U_3O_8. Other details are in the literature (See, e. g., Sherman, 1971, 1972).

Arcot-Sevattur (India).

This deposit, in Madras pyrochlores, contains about 0.06% U_3O_8. The pyrochlores occur in a carbonatite flanked by gneiss and syenite.

Dariba (India).

See Khetri-Dariba.

Jim Jim (Australia).

These uraniferous epigenetic deposits resemble those of Nabarlek (q. v.). See also Rum Jungle.

Katherine-Darwin (Australia).

This area comprises uraniferous deposits of Archean and Proterozoic ages. These are discussed in Chapter 9.

Khetri-Dariba (India).

Uranium occurs in parts of this area of Rajasthan in close association with copper mineralization (See, e. g., Tatsch, 1975a). The mineralization seems to be an extension of the same uraniferous vestiges of seismotectonomagmatic-belt activity seen to the east in the Singhbhum (q. v.) region of Madhya Pradesh. Besides copper and uranium, this belt contains also the Iron-ore series, the Gangpur series, the Sakoli series, and the Aravalli and Delhi systems.

Numerous Cu-Pb-U-Zn mineralizations are seen in the area.

The mineralization areas are primary base-metal and belong to the hypothermal-mesothermal-epithermal types. The asoociated vestiges of seismotectonomagmatic-belt activity are related primarily to a seismotectonomagmatic-belt episode of the late Precambrian that was superimposed on earlier Archean episodes of folding (See, e. g., Tatsch, 1972a). Udas and Mahadevan (1974) have analyzed the controls on the uranium mineralization of that area in terms of (1) the tectonic styles, (2) the metamorphic history, (3) the mineralogy and paragenic sequences in the mineralization, and (4) the relative positions of the locales of the Cu-Pb-Zn with respect to the uranium mineralization. Some of their conclusions may be summarized:

a. Most of the mineralization is syntectonic to late tectonic.

b. Highest uranium values occur in zones of most intense tectonism.

c. The uranium may occur either with or in shear planes parallel to those containing high-temperature assemblages of Cu-Pb-Zn.

d. Mo and Ni concentrations increase with depth.

e. The uranium host rocks are the greenschist to amphibolite facies; higher granulite facies appear to inhibit uranium mineralization.

Nabarlek (Australia).

These uraniferous epigenetic deposits lie within older Precambrian rocks of Northern Territory. They average better than 0.10% U_3O_8 (Sherman, 1972). See also Rum Jungle; Katherine-Darwin. Other details are included in the literature (See, e. g., Langford, 1974).

Palabora (South Africa).

The uranium of this area is associated with the carbonatite complex. Uranothorianite can be recovered as a byproduct of other mining operations in this area. The mineralization is traceable to early Archean episodes of seismotectonomagmatic-belt activity in that area (Tatsch, 1973a).

Rajasthan (India).

See Khetri-Dariba.

Ranger (Australia).

These uraniferous epigenetic deposits resemble those of Nabarlek (q. v.). See also Rum Jungle.

Sarguia (India).

This uranium deposit, in Madhya Pradesh, occurs in a syenite porphyry within granite that appears to be associated with an eroded vestige of Archean seismotectonomagmatic-belt activity (See, e. g., Tatsch, 1973a). The anticline is largely quartzite and marbles. The mineralization contains 0.1% to 0.25% U_3O_8, either as uranium minerals or in fluorite, confined to the matrix of the porphyry.

Singhbhum (India).

Uranium deposits occur within a 160-km-long arcuate vestige of seismotectonomagmatic-belt activity in the Singhbhum thrust belt of south Bihar. This vestige follows the northern margin of the cratonic block of Singhbhum granite. Some of the mineralization is of the hydrothermal vein type. Others is widely disseminated in crush and breccia zones. Better known for its copper deposits (See, e. g., Tatsch, 1975a), the Singhbhum thrust belt also contains apatite-magnetite and kyanite deposits. It trends east-west and traverses Precambrian metasediments comprising closely folded mica schists, quartzite, conglomerates, and metamorphosed basic lavas. Uranium occurs as lenses both along strike and at depth. The main ore minerals are uraninite, torbernite, and autunite. The average grade is 0.067% U_3O_8. The mineralization occurred in two stages: (1) apatite and magnetite, and (2) uranium and the sulfides including chalcopyrite (See, also, Tatsch, 1975a, 1975g). Other details of this deposit are in the literature (See, e. g., Sarkar and Deb, 1974).

<u>Tete</u> (Mozambique).

The davidite deposits at Tete suggest that the uranium was emplaced during a high-temperature hydrothermal phase of Archean seismotectonomagmatic-belt activity in that area (Tatsch, 1973a).

<u>Udaisagar</u> (India).

This deposit lies within the northwestern state of Rajasthan, characterized by Precambrian Aravalli metasediments comprising conglomerates, chloritic and biotitic phyllites, impure limestones, quartzites, black carbonaceous phyllites, and breccias intruded by granites. The uranium mineralization is confined to a kaolinized and brecciated fault gouge within the relatively downthrown western block of the fault striking N20°W. The 6 m by 200 m mineralization zone ranges from 10 m to 30 m below the water table. The uranium occurs as streaks and stringers of uraninite and sooty pitchblende with small amounts of other minerals (e. g., pyrite, siderite, and chalcopyrite) plus clay gouge (68%) and miscellaneous fragments of phyllite, quartz, and unaltered feldspar (32%). Other details of this deposit are included in the literature (See, e. g., Jayaram et al. (1974) and Dar (1972).

Umra (India).

These deposits, located near Udaipur, Rajasthan, contain various secondary uraniferous minerals. These have been produced by oxidation and remobilization of uraninite. Associated with vestiges of seismotectonomagmatic-belt activity, the deposits occur along the lower limbs of anticlines in shear zones within limestones and shales. The source of the mineralization appears to be hydrothermal emanations from Udaisagar granite about 5 km to the northeast. See also Udaisagar.

Conclusions Regarding the Archean Uranium Deposits.

When the known Archean uranium deposits are analyzed, it is found that almost all of them appear to be associated with those areas of the Earth's surface where vestiges of Archean seismotectonomagmatic-belt activity remain today. These are found primarly within the shields of Africa, Asia, Australia, and the Americas.

Chapter 11

THE PRESENT DISTRIBUTION OF URANIUM DEPOSITS AS A FUNCTION OF THE EARTH'S BEHAVIOR DURING THE PAST 4.6 BILLION YEARS

When the global distribution patterns of uranium deposits are analyzed in terms of the Earth's behavior, the first impression is that, although uranium occurs in a limited number of forms, the deposits appear to be found in such a great variety of age provinces as to preclude generalizations. Closer scrutiny reveals, however, that the Earth's present uranium deposits can be interpreted in terms of episodes of geodynamic behavior beginning during the earliest Archean, when the first uranium deposits formed. This can best be seen when the Earth's behavior is viewed in terms of a long-lived, deep-seated driving mechanism such as that provided by the Tectonospheric Earth Model.

Under this concept, the Earth's present uranium deposits are interpreted as vestiges of fossil seismotectonomagmatic betls that have been active one or more times during the Archean, the Proterozoic, or the Phanerozoic. Here it is well to recall that the earliest Archean uranium deposits, under this concept, are derivable from two basic sources: (1) the Earth's tectonosphere, and (2) remnants of Earth Prime (Chapter 2). Some of these original mineral deposits were "re-worked" by the activity associated with the earliest seismotectonomagmatic belts, perhaps as early as 4.6 b. y. ago. Subsequent seismotectonomagmatic-belt activity, during the Archean and through subsequent epochs, including today, has served to: (1) contribute some "new" uranium deposits to the surface of the Earth; and (2) re-work some of the earlier deposits. Vestiges of this activity can be seen in the present uranium deposits of various types in different parts of the world. Some of these deposits may be reviewed briefly to show how their present characteristics show vestiges of the Earth's past behavior.

Uranium in Carbonaceous Shales.

Uranium is associated with Proterozoic carbonaceous shales in many parts of the world, e. g., the Chattanooga shales of the U. S. A. and the South Alligator River complex of Australia. These uraniferous deposits are derivable from the late Archean and Proterozoic vestiges of seismotectonomagmatic-belt activity that occurred in the applicable areas of the Earth's surface (Tatsch, 1972a, 1973a). In the mineralizations, the uranium is strongly associated with Cu, Ga, and V. These minerals are relatable to either the organic fraction or to the minor-element fraction of the shales (See, e. g., Vine and Tourtelot, 1970). But the uranium does not appear to be related to variations in the total carbon content of the shales, because a negative correlation at the 90% level is shown by the data. Thus, although uranium is present in most of these carbonaceous shales (average = 0.0096%), those with high total carbon (organic + mineral) do not necessarily contain high uranium values.

The above and related analyses suggest (1) that the carbonaceous shales have provided a suitable reducing environment for the precipitation of the uranium; (2) that the uranium associates with multivalent Cu and V; but (3) not necessarily that the carbonaceous matter has concentrated the uranium during the deposition (See, e. g., Ayres and Eadington, 1975).

The transport of the uranium appears to have been as uranyl complexes in groundwater (or artesian waters) in the sandstone conglomerate aquifers overlying the unconformity. Wherever underlying shales have been brought into contact with the groundwater by faulting, suitable conditions would have existed for the subsequent reduction and precipitation of uranium. The source of the uranium in the groundwater appears to have been the acid volcanics associated with applicable vestiges of seismotectonomagmatic-belt activity (Tatsch, 1973a).

The Formation of Vein-Type Uranium Deposits.

The details of the origin and evolution of vein-type uranium deposits are not conpletely understood (See, e. g., Barbier, 1974). It is well, however, to summarize what is known about this type of deposit:

a. Vein-type uranium deposits are not as numerous nor as voluminous as are stratified uranium deposits.

b. They are more widespread than are stratified deposits, e. g., western and eastern Europe, North America, and Australia.

c. Uraniferous areas are predominantly associated with granites where geochemical uranium (uraninite) can easily be leached through slight weathering, in the absence of any vegetation, together with Al, Ca, Na, and Si.

d. Aluminum, calcium, and silicon are usually essential constituents of mineral deposits of pitchblende, together with a matrix of calcite, clay, or quartz.

e. Many vein-type deposits appear to have formed during the Permian, when the weathering conditions were favorable for this type of leaching.

f. Veined pitchblende deposits usually contain mineral associations and sequences associated with surficial sources.

Close scrutiny of these basic observations, together with other geometrical, mechanical, thermal, and chemical aspects of vein-type deposits suggests that their formation may be expressed schematically as follows (Tatsch, 1973a):

$$\sum_{i=1}^{n} a_i A_i, \quad n=3, \text{ where}$$

A_1 = hydrothermal process,

A_2 = weathering processes,

A_3 = other processes, and

a_i = constants, any of which may be zero, provided that

$$\sum_{i=1}^{n} a_i = 1.$$

For example, if $i_2 = i_3 = 0$, then $a_1 = 1$; and the deposit was produced completely by hydrothermal processes. If $i_1 = 0.7$ and $i_2 = 0.3$, then $i_3 = 0$; and the deposit was produced by a combination of hydrothermal (70%) and weathering (30%) processes. If $i_1 = 0.7$ and $i_2 = 0.2$, then $i_3 = 0.1$; and the deposit was produced by a combination of hydrothermal (70%), weathering (20%), and other (10%) processes.

Because this scheme permits the definition of an infinity of combinations of processes for the formation of vein-type uranium deposits, all of the Earth's deposits of this type may be defined by this scheme. Thus, according to Barbier (1974), the vein-type deposits associated with the Massif Central of France appear to be best described by the parameters, $i_1 = 0.1$, $i_2 = 0.85$ and $i_3 = 0.05$. Many of the vein-type deposits of North America appear to be best described by the parameters, $i_1 = 0.85$, $i_2 = 0.1$, and $i_3 = 0.05$. Appropriate parameters may be determined for other vein-type deposits. When this is done, it is found that no two vein-type uranium deposits are exactly alike.

If the Earth's vein-type uranium deposits have one thing in common, it is the "origin" of the uranium they now contain. In every case, the "ultimate" source must be the uranium that was present in the Earth during the "earliest Archean" (i. e., roughly

4.6 b. y. ago). Exactly how this uranium was distributed at that time is not completely understood (See, e. g., Tatsch, 1973a). All that is known is that the present uranium deposits are vestiges of the uranium that the Earth contained 4.6 b. y. ago. The differences among the present deposits reflect the spatio-temporal differences that have existed within the geometrical, mechanical, thermal, and chemical aspects of the "re-working" (including weathering) that the Earth's original uranium has undergone during the past 4.6 b. y. to produce today's uranium deposits. One appraoch for analyzing these differences is provided by the Tectonospheric Earth Model (Tatsch, 1972a, 1973a). This concept explains, for example, why some uranium deposits have undergone very little "re-working" since the Archean, while others have been affected by episodes as recently as the Cenozoic. Some of these processes are still affecting some of today's deposits (Chapter 6).

Uranium in Coal Formations.

The uraniferous deposits associated with coal formations are vestiges of the same episodes of seismotectonomagmatic-belt activity that were instrumental in the formation of the coal basins (Tatsch, 1972a; Danchev and Strelyanov, 1973). Genetically, the associations are of three types: (1) diagenetic, with the uraniferous deposits formed during peat-formation stages; (2) epigenetic, with the uraniferous deposits of later origin; and (3) combinations formed during several stages of lithogenesis.

Uranium in Sea Water.

Sea water contains roughly 1 ppb (part per billion), or 0.0000001%, of uranium. Although this is a very low concentration of uranium, the size of the oceans suggests that this is a possible source of uranium. This is particularly true in connection with desalination and magnesium-extracting processes.

Marine phosphorite is the dominant source of phosphate and constitutes a very large source of uranium (See, e. g., Bone Valley in Chapter 6, and Phosphoria in Chapter 8). In these rocks, the vestiges of the younger seismotectonomagmatic belts (i. e., geosynclines) tend to be more uraniferous than are the older vestiges such as cratons (Tatsch, 1973a). The Phosphoria formation is a vestige of Permian seismotectonomagmatic-belt activity in Idaho, Montana, Utah, and Wyoming. The Bone Valley formation is a vestige of Pliocene seismotectonomagmatic-belt activity in Florida.

Other areas of hhe world contain similar vestiges of various ages. Large portions of these vestiges are uraniferous along the Mediterranean Sea from Morocco to Israel. Near Recife, Brazil, deposits reminiscent of those in Bone Valley, Florida, yield phosphates containing 0.02% U_3O_8. Phosphorites fill large depressed vestiges of seismotectonomagmatic-belt activity in the Central African Republic. The aluminous phosphates in Senegal and Nigeria are also uraniferous. The marine phosphorites in the Kara-Tau Mountains, Kazakhstan, lie within vestiges of a Cambrian episode of seismotectonomagmatic-belt activity. Four separate basins on the continental sheld of South West Africa, including one near Walvis Bay (q. v.) are uraniferous vestiges of seismotectonomagmatic-belt activity. Similar uraniferous muds are found in most other seas (e. g., Baltic, Black, and Caspian, where there are concentrations of 0.001% U_3O_8).

Uranium Deposits Associated with Hydrocarbons.

Pitchblende and some other uraniferous ores are associated with hydrocarbons in various parts of the world. This may occur within vein-type and stratiform deposits. These ore beds appear to have evolved within vestiges of seismotectonomagmatic-belt activity in three main stages:

a. The deposition of bitumens in favorable structural and lithological conditions remaining as vestiges of earlier seismo-tectonomagmatic-belt activity (Tatsch, 1974b).

b. The precipitation of uranium from the circulating solutions by bitumens.

c. The carbonization and oxidation of the bitumens, accompanied by the recrystallization and redistribution of uranium during the alteration process.

In most of the uranium-and-hydrocarbon associations, the hydrocarbons appear to have polymerized from a gaseous phase, through a liquid phase, to a solid. In most cases, the uranium is in oxide form, usually scattered as sub-microscopic inclusions of pitchblende. In the uranium-rich bitumens, there are large deposits of pitchblende together with the smaller inclusions. In some cases, the pitchblende inclusions occur within the cement of sandstones and siltstones and within the shrinkage cracks of the bitumens. Most of the pitchblende appears to have been associated with later formations.

A large part of the uranium appears to have been associated with bitumen prior to polymerization, with the bitumen acting as ion-exchange resins (See, e. g., Kornechuk and Burtek, 1974). A process of uranium sorption and desorption occurred after the oxidation of the bitumen increased its carboxyl (COOH) group and, thereby, its sorptive capacity. During subsequent carbonization of the bitumen, the uranium separated out as oxides. This segregation of the uranium within the bitumen suggests repeated recrystallization and redistribution of the uranium during the polymerization of the bitumen.

Uranium in Brines.

Small amounts of uranium occur in brines. This is true of closed basins on all continents and has probably occurred throughout geologic time. The heaviest concentrations seem to occur within some oil-feild brines. This suggests that petroleum is capable of transporting small amounts of uranium, but this is not an important source of uranium.

Many asphalt-bearing rocks are uraniferous, with concentrations up to 0.02% U_3O_8. Host rocks in which the asphalt occurs include arkose, diatomite, limestone, and tuff. The uranium occurs as an organo-uranium complex within the asphalt, rather than in the host rock. This suggests that the uranium occurred as an original constituent of the oil or was introduced during migration (See, e. g., Von Backström, 1974b; Tatsch, 1974b).

The Formation of Roll-Type Uranium Deposits.

Roll-type uranium deposits occur in sandstone vestiges of seismotectonomagmatic-belt activity that contain pyrite and other reduced minerals (Tatsch, 1973a, 1975g). The deposits appear to have been formed by ore-bearing solutions that contained oxygen and moved in the plane of the sandstone bed (See, e. g., Warren, 1972). Over a long period of time, the ground-water oxygen appears to have destroyed most of the pyrite, particularly in upstream, altered zones, with reposition farther downstream.

Chemical disproportionation appears to have been the principal means by which some roll-type deposits were produced (e. g., Shirley Basin). Other details are included in the individual discussions of roll-type deposits in earlier chapters.

Uranium Deposits Produced by Weathering.

Most ore minerals are formed at temperatures higher than atmospheric and in reducing environments. Wherever these minerals are exposed to surface conditions, they break down chemically, thereby forming new compounds or going into solution. Most rock-forming minerals are unstable at the Earth's surface. This causes them to undergo chemical changes as equilibrium is re-established with the environment. Minor amounts of metals within unweathered rock may be concentrated into economic deposits during weathering. This may be seen particularly after each episode of seismotectonomagmatic-belt activity that has occurred during the past 3.6 b. y. (See, e. g., Tatsch, 1973a). Enrichment normally occurs wherever the vestiges of the seismotectonomagmatic-belt activity contain stable, oxidized metallic products or wherever other constituents are selectively leached away.

All minerals are soluble under favorable conditions. Their solubilities vary over a wide range. Consequently, the rates of chemical reactions vary independently, and there is a wide range of effective mineral stabilities. Meteoric waters charged with CO_2 and O_2 from the atmosphere will carbonatize, hydrate, and oxidize the rock-forming minerals. Many sulfides are converted to sulfates, most of which are soluble. Others are converted to the more-stable carbonates, oxides, and native metals. Thus, aluminim, iron, and manganese form relatively insoluble hydroxides and oxides (Tatsch, 1975g). Lead forms a stable sulfate. Copper, lead, and zinc form carbonates in many environments produced by seismotectonomagmatic-belt activity (Tatsch, 1975a). Other metals are retained as silicates, e. g., other copper, chromium, nickel, and zinc. Some copper, silver, and gold remain as native metals, depending upon the exact nature of the seismotectonomagmatic-belt activity involved (Tatsch, 1973a).

The mobility of metal ions in the weathering zone is a function primarily of the specific compositions of the vadose waters and country rocks that are found within the particular vestige of seismotectonomagmatic-belt activity being analyzed (Tatsch, 1973a).

Thus, sulfide-free meteoric waters can be expected to leach certain metals from the igneous rocks within a vestige of seimsotectonomagmatic-belt activity. This is the basis for the formation of uranium deposits as a result of weathering. The sulfide-free waters leach uranium (and some other metals such as zinc and molybdenum) from the igneous rocks, thereby leaving residues of stable oxidation products of aluminum, chromium, iron, titanium, and some other oxides.

Uranium Deposits in Metamorphic Rocks.

Many metamorphic rocks contain concentrations of uranium. The exact mechanism for the emplacement of these deposits is metamorphic rocks is not known. An initial analysis suggests, however, that the metamorphism serves to "re-work" previously existing uranium (Tatsch, 1973a). This may be seen in all episodes of metamorphism.

For example, during the metamorphism associated with the Alpine seimmotectonomagmatic-belt activity, some of the Permian uranium and associated non-ferrous metal deposits of many parts of the world were mobilized and underwent complex recrystallization. This casued some of the older (Permian) ores to migrate to neighboring rocks. Examples of this may be seen in all contientts, including particularly the deposits at St. Bernhard (q. v.) and Preit Valley (q. v.).

As interpreted by the seismotectonomagmatic-belt concept, the same mechanism that causes the metamorphism also causes the possible addition of "new" uranium from subsurface sources (Tatsch, 1973a).

Uranium Provinces as Vestiges of Seismotectonomagmatic-Belt Activity during the Past 4.6 Billion Years: A Summary.

There are three types of environments or provinces with which uranium deposits appear to have been associated during the past 4.6 b. y.:

a. Unmodified seismotectonomagmatic belts. The uranium in presently unmodified seismotectonomagmatic belts is derivable from two sources: (1) the Earth's mantle, and (2) re-worked vestiges of older seismotectonomagmatic belts (Tatsch, 1973a).

b. Sedimentary areas. The urainum in the Earth's sedimentary areas is derivable from the vestiges of uraniferous seismotectonomagmatic belts. The typical geological setting for these areas includes a sequence of up to 3.5 km of sediments, no more than gently folded, and unconformably overlying highly distorted ancient basement rocks (Tatsch, 1973a; Dunham, 1974). Block faulting and related vestiges of seismotectonomagmatic-belt activity may or may not be present. A simple suite of elements besides uranium reaches economic concentrations in this setting, e. g., Ba, Cu, F, Pb, and Zn. These economic deposits appear to be attributable primarily to three aspects of seismotectonomagmatic-belt activity: (1) the uraniferous vestiges of earlier seismotectonomagmatic-belt activity; (2) the permeable paths of preferential flow provided within the seismotectonomagmatic-belt vestiges; nnd (3) the fissure system associated with seismotectonomatmatic-belt activity.

c. Precambrian cratonic provinces. The uranium in the Earth's Archean and Proterozoic cratons is derivable from a combination of the processes described in a and b, above (Tatsch, 1973a).

Chapter 12

THE TECTONOSPHERIC EARTH MODEL AS A SUPPLEMENTARY TOOL IN THE EXPLORATION FOR URANIUM DEPOSITS

An analysis of the Earth's uranium deposits shows that they are not homogeneously distributed. According to the Tectonospheric Earth Model, they should be distributed within certain structural elements of the Earth's crust that are associated with vestiges of seismotectonomagmatic-belt activity (Tatsch, 1973a). This is particularly true of certain structures that have been produced by sialic phases of seismotectonomagmatic-belt magmatism. This preferential distribution applies both to the original emplacement of the uranium deposits and to their enrichment through hypogenic and supergenic remobilizations associated with subsequent episodes of seismotectonomagmatic-belt activity.

This suggests that the seismotectonomagmatic-belt concept should provide a valuable tool in the exploration for hidden uranium deposits.

The Seismotectonomagmatic-Belt Concept Applied to the Exploration for Uranium Deposits.

Because most uranium deposits that display easily recognizable surficial manifestations have already been found, the future success in uranium exploration will depend increasingly upon a more sophisticated approach, preferably one that embodies a clear understanding of the genetic relationships that exist between mineral deposits and geodynamic processes and phenomena. One such approach has been used in this book: it is based on the observation that economic concentrations of minerals are not randomly distributed but appear to be related to tectonic processes and regimes within the Earth. The details of these genetic relations are not known, and speculation regarding them has long been the subject of discussions among geologists. The resulting diversity of opinion can be seen by reading any applicable journal or by attending any meeting devoted to economic geology.

In the first 11 chapters of this book, an attempt has been made to determine how the Earth's uranium deposits have originated, have evolved, and have become emplaced into thier present distribution patterns in accordance with a single, long-lived, deep-seated, global mechanism that has eeen operating within the Earth during the 4.6 b. y. that the Earth is believed to have been in existence. This mechanism, the Tectonospheric Earth Model, permits the prediction of certain variations that have occurred within the geometrical, mechanical, thermal, and chemical heterogeneities of the Earth's upper 1000 km during the past 4.6 b. y. The ability of the model to provide these predictions forms the basis for its use as a supplementary tool in the exploration for uranium deposits.

The Tectonospheric Earth Model as a Research Tool in Defining Uraniferous Basins.

As one example of its use as a supplementary tool in the exploration for uranium ddposits, the Tectonospheric Earth Model may be used as a guide in defining the sedimentary framework of basins that might be expected to contain uranium deposits in sandstone formations. Some of the desirable conditions that the Tectonospheric Earth Model concept can define on the basis of interpreting vestiges of seismotectonomagmatic-belt activity may be listed:

a. The original boundaries and configuration of basins, particularly those intraformational basins within thick sequences of rocks.

b. The location, relief, and rock types of bordering highlands.

c. The diagnostic characteristics of the associated rock facies, particularly those between the centers and margins of the basins.

d. The present and erstwhile hydrolic patterns expected within the predicted vestiges of seismotectonomagmatic-belt activity.

e. The post-sedimentary histories indigenous to the applicable seismotectonomagmatic-belt episodes.

f. The origin, evolution, and present characteristics of different types of formations in which uranium is known to be concentrated and how these are related to specific types of seismotectonomagmatic-belt activity.

g. The cause for the differences (and similarities) among various types of uranium deposits; for example, how specific roll and tabular types of deposits are related to each type of seismotectonomagmatic-belt activity (Tatsch, 1973a). More specifically, this concept shows why much of the uranium in some areas favors roll-type deposits (e. g., that of the Colorado Plateau), whereas much of that in other areas favors tabular-type deposits (e. g., that of the Wyoming Tertiary basins).

h. The reason why two or more types of uranium deposits form in some areas but not in others (e. g., tabular-type sediments in the Tertiary of Wyoming and other types of deposits in older vestiges of seismotectonomagmatic-belt activity in that area).

i. How the geology ond distribution patterns of large uraniferous veins are related to patterns of vestiges of uraniferous seismotectonomagmatic-belt activity.

Favorable Targets for New Uranium Deposits.

The most favorable prospects for exploration for new uranium deposits are those that are identifiable with vestiges of seismotectonomagmatic-belt activity that: (1) contained part of the "original" Archean uranium of the Earth's tectonosphere; and (2) added "new" uranium from subsurface sources (Tatsch, 1973a). These vestiges include:

a. The margins and certain axial parts of sedimentary basins, particularly those containing sediments derived from specific types of seismotectonomagmatic-belt activity that produce granitic terranes and volcanic ash. These basins may be either intermontane (e. g., Wyoming) or gulf (e. g., Texas) type, provided they are identifiable with vestiges of uranifersou seismotectonomagmatic-belt activity (Tatsch, 1973a). Most of these sedimentary-type uraniferous deposits can be expected within the Tertiary and Cretaceous. But, wherever suitable undisturbed environments have been provided, this type of sedimentary uraniferous ores should be found also in older sediments that are identifiable with vestiges of seismotectonomagmatic-belt activity.

b. Some Mesozoic and Paleozoic uraniferous deposits may be expected within unconformities related to paleokarst formations. This is particularly true of two types of paleokarst structures:

(1). Cretaceous uraniferous phosphatic rock that fills paleokarst developed within Precambrain dolomite. An example of this is Bakouma, Central African Republic.

(2). Uraniferous deposits of tyuyamunite-bearing breccias that fill the cavities related to karst structures. An example of this is the Madison Limestone of the Pryor Mountains, Montana.

c. Uraniferous Proterozoic quartz-pebble conglomerates, identifiable with Archean and early Proterozoic seismotectonomagmatic-belt activity, should be found on all continents. Although some continents do not have outcrops of uraniferous quartz-pebble conglomerates, many promising deposits of this type should be present, however, within shallow depths beneath the surface. This is particularly true, for this specific type of deposit, where the last major episode of seismotectonomagmatic-belt activity occurred during the early Proterozoic (say 2.5 b. y. ago). Examples of outcrops of these old uraniferous quartz-pebble conglomerates are those at Witwatersrand and at the Blind River - Elliot Lake area. The same type of deposits should exist below the surface in all continents, e. g., the vestiges of major, Proterozoic seismotectonomagmatic-belt activity in the Dakotas and eastern part of Montana (U. S. A.) about 2.6 b. y. ago (See, e. g., Goldich,et al., 1966).

d. A close scrutiny of the origin, evolution, and present characteristics of mineral deposits suggests that significant quantities of uraniferous ores should be found within porphyry-type uranium deposits (Tatsch, 1973a). These should resemble the well-known prophyry-type molybdenum and copper deposits in various parts of the world (Tatsch, 1975a). Theoretical considerations and observational evidence suggest that porphyry uranium deposits would be associated with late magmatic differentiates of seismotectonomagmatic-belt activity. Thus, the uraniferous porphyries might be associated with alaskites and pegmatites, particularly within the arid regions of the Earth. The Rössing (q. v.) deposit of South West Africa might serve as a representative example of this type.

These four targets are representative, rather than exhaustive, of the approaches that may be derived from the Tectonospheric Earth Model as a supplementary tool in the exploration for uranium deposits.

Others have been suggested in the first eleven chapters of this book. Still others will be obvious after a careful consideration and application of the principles embodied in the concepts regarding the origin, evolution, and present characteristics of the Earth's seismotectonomagmatic belts and the uranium deposits associated therewith in specific areas of interest to the reader.

REFERENCES

Adams, J. K., and Weeks, A. M., 1974. Paleoenvironmental control of uranium deposits in South Texas. Econ. Geol., 69, 1175 (Abs).

Adams, S. S., Curtis, H. S., and Hafen, P. L., 1974. Alteration of detrital magnetite-ilmenite in continental sandstones of the Morrison Formation, New Mexico. In: IAEA (q. v.), pp. 219-252.

Adler, H. H., 1970. Interpretation of color relations in sandstone as a guide to uranium exploration and ore genesis. In: IAEA (q. v.), pp. 331-344.

Adler, H. H., 1972. Exploration for uranium in sandstones: Geochemical, remanent magnetics,and sulfur-isotope applications. In: IAEA (q. v.).

Adler, H. H., 1974. Concepts of uranium-ore formation in reducing environments in sandstones and other sediments. In: IAEA (q. v.), pp. 141-168.

Altschuler, Z. S., Jaffe, E. B., and Cuttitta, F., 1956. The aluminum phosphate zone of the Bone Valley Formation, Florida, and its uranium deposits. U. S. Geol. Survey Ppr. 300, 483-387.

Amerigian, C., 1974. Seafloor dynamic processes as the possible cause of correlations between paleoclimatic and paleomagnetic indices in deep-sea sedimentary cores. Earth Planet. Sci. Ltr., 21, 321-326.

Amstutz, G. C., and Bernard, A. J. (Editors), 1973. Ores in Sediments. Springer-Verlag, Berlin, 350 pp.

Anderson, R. N., 1974. Cenozoic motion of the Cocos plate relative to the asthenosphere and cold spots. Geol. Soc. Amer. Bull., 85, 175-180.

Anderson, D. L., 1975. Chemical plumes in the mantle. Geol. Soc. Amer. Bull., 86, 1593-1600.

Anderson, T. A., 1975. Carboniferous subduction complex in the

Harz Mountains, Germany. Geol. Soc. Amer. Bull., 86, 77-82.

Anderson, R. Y., and Kirkland, D. W., 1960. Orgin, varves, and cycles of Jurassic Todilto Formation, New Mexico. Bull. Amer. Assoc. Petrol. Geol., 44, 37-52.

Andrade-Ramos. See De Andrade-Ramos.

Anguita, F., and Hernan, F., 1975. A propagating fracture model versus a hot spot origin for the Canary Islands. Earth Planet. Sci. Ltr., 27, 11-19.

Anhaeusser, C. R., Mason, R., Viljoen, M. J., and Viljoen, R. P., 1969. A reappraisal of some aspects of Precambrian shield geology. Bull. Geol. Soc. Amer., 80, 2175-2200.

Anon., 1971. India: Recent mineral discoveries. Min. J., 277, 507.

Anon., 1975a. Increased reserves make Jabiluka Australia's largest single uranium deposit. Eng. Min. J., 176(1), 37.

Anon., 1975b. Seven-state search for new U. S. uranium resources will be funded by USGS. Eng. Min. J., 176(7), 128.

ARCYANA (Francheteau, J., and 7 others), 1975. Transform fault and rift valley from bathyscaph and diving saucer. Science, 190, 108-116.

Armstrong, F. C., 1974a. Estimation of uranium resources ultimately recoverable from Gas Hills district, central Wyoming. Econ. Geol., 69, 149 (Abs).

Armstrong, F. C., 1974b. Uranium resources of the future: "porphyry" uranium deposits. In: IAEA (q. v.), pp. 625-635.

Ashgirei, G. D., 1974. Origin of continental crust. Geology, 2, 401-404.

Augustithis, S. S., Mposkos, E., and Vgenopoulos, A., 1974. Geochemical and mineralogical studies of euxinite and its alteration products in graphitic pegmatites from Harrar, Ethiopia. Proc. Symp. Form. Uran. Ore Dep., 61-71.

Aumento, F., 1971. Uranium content of mid-ocean basalts.

Earth Planet. Sci. Ltr., 11, 90-92.

Aumento, F., and Hyndman, R. D., 1971. Uranium content of the oceanic upper mantle. Earth Planet. Sci. Ltr., 12, 373-382.

Austin, P. M., 1975. Paleogeographic and paleotectonic models for the New Zealand geosyncline in eastern Gondwanaland. Geol. Soc. Amer. Bull., 86, 1230-1234.

Ayres, D. E., and Eadington, P. J., 1975. Uranium mineralization in the South Alligator River Valley. Min. Dep., 10, 27-42.

Azhgirei, G. D., 1968. Folding and mountain building (orogenesis). In: Mistik (Editor) (q. v.), pp. 287-298.

Backström. See von Backström.

Badham, J. P. N., Robinson, B. W., and Morton, R. D., 1972. The geology and genesis of the Great Bear Lake silver deposits. In: Gill (Editor) (q. v.), pp. 541-548.

Bain, G. W., 1968. Syngenesis and epigenesis of ores in layered rocks. In: Stemprok (Editor) (q. v.), pp. 119-136.

Bak, J., Korstgård, J., and Sørensen, K., 1975. A major shear zone within the Nagssugtoqidian of West Greenland. Tectonophys., 27, 191-209.

Baker, B. H., Mohr, P. A., and Williams, L. A. J., 1972. Geology of the Eastern Rift System of Africa. Geol. Soc. Amer., Boulder, 67 pp.

Ballard, R. D., Bryan, W. B., Heirtzler, J. R., Keller, G., Moore, J. G., and Van Andel, Tj., 1975. Manned submersible observations of the FAMOUS area: Mid-Atlantic Ridge. Science, 190, 103-108.

Banerjee, P. K., and Gosh, S., 1972. Correlation of Precambrian ore provinces of east Africa, India, and West Australia. Econ. Geol., 67, 55-62.

Barbey, P., 1975. The Adam Talka epizonal sequence (North Mauritania) and a general scheme for the Eburnean orogenic belt of west Africa. Precamb. Res., 2, 255-262.

Barbier, M. J., 1974. Continental weathering an a possible origin of vein-type uranium deposits. Min. Dep., 9, 271-288.

Barnes, F. Q., and Ruzicka, V., 1972. A genetic classification of uranium deposits. In: Gill (Editor) (q. v.), pp. 159-166.

Barthel, F. H., 1974. Review of uranium occurrences in Permian sediments in Europe, with special reference of uranium mineralization in Permian sandstone. In: IAEA (q. v.), pp. 277-289.

Barton, J. M., Jr., 1975. Rb-Sr isotopic characteristics and chemistry of the 3.6 b. y. Hebron gneiss, Labrador. Earth Planet. Sci. Ltr., 27, 427-435.

Basham, I. R., and Rice, C. M., 1974. Uranium mineralization in Siwalik sandstones from Pakistan. In: IAEA (q. v.), pp. 405-418.

Bateman, A. M., 1950. Economic Mineral Deposits, 2nd ed. Wiley, New York, 916 pp.

Bateman, A. M., 1951. The Formation of Mineral Deposits. Wiley, New York, 371 pp.

Baturin, G. N., 1974. Uranium in the modern sedimentary cycle. Geochem. Internat., 10, 1021-1030.

Becraft, G. E., and Weis, P. L., 1963. Geology and mineral deposits of the Turtle Lake quadrangle, Washington. U. S. Geol. Surv. Bull. 1131, 73 pp.

Berg, E., and Sutton, G. H., 1975. Dynamic interaction of seismic and volcanic activity of the Nazca plate edges. Phys. Earth Planet. Int., 9, 175-182.

Berger, I. A., 1974. The role of organic matter in the accumulation of uranium. In: IAEA (q. v.), pp. 99-124.

Berzina, I. G., Yelisseyeva, O. P., and Popenko, D. P., 1973. Control of uranium distribution within the intrusive rocks in north Kazakhstan. Proc. USSR Acad. Sci., Geol., 73(7), 16-25.

Berzina, I. G., Yeliseyeva, O. P., and Popenko, D. P., 1974. Distribution relationships of uranium in intrusive rocks of northern Kazakhstan. Internat. Geol. Rev., 16, 1191-1204.

Best, M. G., 1975. Migration of hydrous fluids in the upper mantle and potassium variation in calc-alkaline rocks. Geology, 3, 429-432.

Bickford, M. E., and Van Schmus, W. R., 1973. Possible middle and late Precambrian igneous arcs in the mid-continent region of North America. Geol. Soc. Amer. Abstr. Prog., 5, 300.

Bickle, M. J., Martin, A., and Nisbet, E. G., 1975. Basaltic and peridotitic komatiites and stromatolites above a basal unconformity in the Belingwe greenstone belt, Rhodesia. Earth Planet. Sci. Ltr., 27, 155-162.

Bigotte, G., and Obelliannne, J. M., 1968. Decouverte de mineralisations unraniferes au Niger. Min. Dep., 3, 317-333.

Bird, P., and Phillips, J. D., 1975. Oblique spreading near the Oceanographer fracture. J. Geophys. Res., 80, 4021-4027.

Black, L. P., Moorbath, S., Pankhurst, R. J., and Windley, B. F., 1973. ^{207}PB / ^{206}Pb whole rock age of the Archaean granulite facies metamorphic event in West Greenland. Nature, Phys. Sci., 244, 50-51.

Blatt, H., and Jones, R. L., 1975. Proportions of igneous, metamorphic, and sedimentary rocks. Geol. Soc. Amer. Bull., 86, 1085-1088.

Bogush, I. A., and Savchenko, N. A., 1971. Rhythmic stratified ores of the Vlasenchikhino copper-pyrite deposit in the north Caucasus. Geol. Ore Dep., 13(3), 94-99.

Bohse, H., Rose-Hansen, J., Sorensen, H., Steenfelt, A., Lovborg, L., and Kunzendorf, H., 1974. On the behavior of uranium during crystallization of magmas, with special emphasis on alkaline magmas. Proc. Symp. Form. Uran. Ore Dep., 49-60.

Bonatti, E., Zerbi, M., Kay, R., and Rydell, H., 1976. Metalliferous deposits from Apennine ophiolites: Mesozoic equivalents of modern deposits from ocean spreading centers. Geol. Soc. Amer. Bull., 87, 83-94.

Bondam, J., and Sorensen, H., 1958. Uraniferous nepheline syenites and related rocks in the Ilimaussaq area, Julianehaab District, Southwest Greenland. Proc. Internat. Conf. Pfl. Uses Atom En., 2, 555-559.

Bondi, A., Carrara, C., and Polizzano, C., 1973. Uranium and heavy metals in Permian sandstones near Bolzano (northern Italy). In: Amstutz and Bernard (Editors) (q. v.), 65-71.

Borrello, A. J., 1972. The Precordillera as a type of geosyncline in Argentina. Internat. Geol. Cong. Abstr., 24, 67.

Bossi, J., 1974. Use of fluid inclusions in the genetic study of vein deposits. In: IAEA (q. v.), pp. 583-592.

Bostrom, K., Farquharson, B., and Eyl., W., 1972. Submarine hot springs as a source of active ridge sediments. Chem. Geol., 10, 189-198.

Bourrel, J., and Pfiffelman, J. P., 1972. La province uranifere du bassin de Franceville (Republique Gabonaise). Min. Dep., 7, 323.

Bowie, S. H. U., 1970. World uranium deposits. In: IAEA (q. v.), pp. 23-33.

Bowie, S. H. U., Davis, M., and Ostle, D. (Editors), 1972. Uranium Prospecting Handbook. Inst. Min. Metall., London, 346 pp.

Bowin, C., 1974. Migration of a pattern of plate motion. Earth Planet. Sci.,Ltr., 21, 400-404.

Bridgwater, D., and McGregor, V. R., 1974. Field work on the very early Precambrian rocks of the Isua area, southern West Greenland. Grøn. Geol. Undersøg. Rapp., 65, 49-59.

Bridgwater, D., Watson, T., and Windley, B. F., 1973. The Archean craton of the North Atlantic region. Phil. Trans. Roy. Soc. Lond., A-273, 493-501.

Brinck, J. W., 1974. The geochemical distribution of uranium as a primary criterion for the formation of ore deposits. Proc. Symp. Form. Uran. Ore Dep., 21-32.

Brobst, D. A., and Pratt, W. P. (Editors), 1973. United States

Mineral Resources. Govt. Prtg. Off., Washington, 722 pp.

Brookins, D. G., 1975. Coffinite-uraninite stability relations in Grants mineral belt, New Mexico (Abs). Amer. Assoc. Petrol. Beol. Bull., 59, 905.

Brookins, D. G., and Lee, M. J., 1974. Clay mineral: uranium mineralization: organic carbonaceous matter relationship in the Grants mineral belt, New Mexico. Econ. Geol., 69, 1177 (Abs).

Brooks, R. A., 1975. Stratigraphic control of uranium mineralization in western Karnes Coutnty, Texas (Abs). Amer. Assoc. Petrol. Geol. Bull., 59, 905-906.

Brown, W. G., 1975. Casper Mountain area (Wyoming): Structural model of Laramide deformation (Abs). Amer. Assoc. Petrol. Geol. Bull., 59, 906.

Bulakh, A. G., Mazalov, A. A., Saturin, A. A., and Bakhtiarov, A. V., 1974. Distribution of uranium and thorium in the alkalic rocks of the Turiy peninsula (Murmansk region). Geochem. Internat., 10, 1063-1065.

Bullen, K. E., 1974a. Introductory remarks on standard Earth model. Phys. Earth Planet Int., 9, 1-3.

Bullen, K. E., 1974b. Standard Earth model requirements in respect to density and rigidity in the inner core. Phys. Earth Planet. Int., 9, 41-44.

Burrett, C. F., 1974. Plate tectonics and the fusion of Asia. Earth Planet. Sci. Ltr., 21, 181-189.

Butler, A. P., Jr., 1974. Some unsettled aspects of potential uranium resources (Abs). Econ. Geol., 69, 149.

Butler, A. P., Jr., Finch, W. I., and Twenhofel, W. S., 1962. Epigenetic uranium deposits in the U. S., exclusive of Alaska and Hawaii. U. S. Geol. Surv. Min. Inv. Res. Map MR-21, 42 pp.

Buzzalini, A. D., and Gloyn, R. W., 1972. The geology and origin of uranium deposits in the Shirley Basin, Wyoming. In: Gill (Editor) (q. v.), pp. 423.

Cabre, R., 1975. Geodynamics in the eastern Pacific and the western Americas. Phys. Earth Planet. Int., 9, 169-173.

Cadigan, R. A., and Felmlee, J. K., 1975. Radioactive mineral springs in Delta County, Colorado (Abs). Econ. Geol., 70, 1318.

Cannon, H. L., 1957. Description of indicator plants and methods of botanical prospecting for uranium deposits on the Colorado Plateau. U. S. Geol. Surv. Bull. 1030-M.

Cauthorn, R. G., 1975. Degrees of melting in mantle diapirs and the origin of ultrabasic liquids. Earth Planet. Sci. Ltr., 27, 113-120.

Cepeda, J. C., 1973. Evidence for multiple Precambrian deformation in the Sangre de Cristo Mountains, New Mexico. Geol. Soc. Amer. Abstr. Prog., 5, 470.

Challinor, J., 1964. A Dictionary of Geology, 2d ed. Oxford Univ. Press, New York, 289 pp.

Challis, G. A., 1975. Pyrite-haematite alteration as a source of colour in red beds and regolith. Nature, 255(5508), 471.

Chapman, D. S., and Pollack, H. N., 1975. Global heat flow: A new look. Earth Planet. Sci. Ltr., 28, 23-32.

Chappell, J., 1974. Upper mantle rheology in a tectonic region: Evidence from New Guinea. J. Geophys. Res., 79, 390-398.

Chase, C. G., and Gilmer, T. H., 1973. Precambrian plate tectonics: The midcontinent gravity high. Earth Planet. Sci. Ltr., 21, 70-78.

Chenoweth, W. L., 1975. Uranium deposits of Nacimiento-Jemez region, Sandoval and Rio Arriba counties, New Mexico (Abs). Amer. Assoc. Petrol. Geol. Bull., 59, 907.

Clarke, W. B., and Kugler, G., 1973. Dissolved helium in groundwater: A possible method for uranium and thorium prospecting. Econ. Geol., 68, 243-251.

Coffin, R. C., 1921. Radium, uranium, and vanadium deposits of southwestern Colorado. Colo. Geol. Surv. Bull. 16.

Compston, W., and McElhinny, M. W., 1975. The Rb-Sr age of the Mashonaland dolerites of Rhodesia and its significance for paleomagnetic correlation in southern Africa. Precamb. Res., 2, 305-315.

Corliss, J. B., 1973. Volcanism and ore-formation related to the subduction of ocaanic crust: the geostill concept. Geol. Soc. Amer. Abstr. Prog., 5, 26,

Corniciuc. See Kornechuk.

Crawford, E., 1975. Developers eye Texas potential for in-situ uranium leaching. Eng. Min. J., 176(7), 81-82.

Crook, K. A., 1974. Petrology of Parry Group, Upper Devonian - Lower Carboniferous, Tamworth-Nundle district, New South Wales. J. Sed. Petrol., 30, 538-552.

Dahl, A. R., and Hagmaier, J. L., 1974. Genesis and characteristics of the southern Powder River Basin uranium deposit, Wyoming, U. S. A. In: IAEA (q. v.), pp. 201-218.

Dall'aglio, M., Gragnani, R., and Locardi, E., 1974. Geochemical factors controlling the formation of the secondary minerals of uranium. Proc. Symp. Form. Uran. Ore Dep., 33-48.

Dalrymple, G. B., Jarrard, R. D., and Clague, D. A., 1975. K-Ar ages of some volcanic rocks from the Cook and Austral Islands. Geol. Soc. Mmer. Bull., 86, 1463-1467.

Damon, P. E., 1972. Epigenic-epeirogenic periodicity. Internat. Geol. Cong. Astrr., 24, 72.

Danchev, V. I., and Strelyanov, N. P., 1973. The main uranium deposits associated with coal formations. Geol. Ore Dep., 15(3), 66-81.

Danchev, V. I., Iliev, P. D., and Lapinskaya, T. A., 1969. Method of studying the exogenic uranium deposits associated with terrigenous rocks. Geol. Ore Dep., 11(5), 34-47.

Dandurand, J. L., Fortune, J. P., Perami, R., Schott, J., and Tollon, F., 1972. On the importance of mechanical action and thermal gradient in the formation of metal-bearing deposits. Min. Dep., 7, 339-350.

Dar, K. K., 1972. Geological environments of uranium and thorium deposits in India: In: Gill (Editor) (q. v.), pp. 167-171.

Dauvillier, A., 1956. L'Origine des Planetes. Presses Universitaires de France, Paris, 225 pp.

Dauvillier, A., 1963. L'Origine du systeme terre-lune. Bull. Soc. Roy. Sci. Liege, 32, 38-53.

Davidson, C. F., and Atkin, D., 1953. On the occurrence of uranium in phosphate rock. Internat. Geol. Cong., 19(11), 13-31.

Davis, J. D., and Guilbert, J. M., 1973. Distribution of the radioelements potassium, uranium, and thorium in selected porphyry copper deposits. Econ. Geol., 68, 145-160.

De Andrade-Ramos, J. R., and Fraenkel, M. O., 1974. Uranium occurrences in Brazil. In: IAEA (q. v.), pp. 637-658.

De Nault, K. J., 1975. Model for formation of sandstone-type uranium deposits in Wyoming with suggestions for exploration (Abs). Amer. Assoc Petrol. Geol. Bull., 59, 907-908.

Denson, N. M., and Gill, J. R., 1956. Uranium-bearing lignite and its relation to volcanic tuffs in eastern Montana and North and South Dakota. U. S. Geol. Surv. Prof. Ppr. 300, 413-418.

Dewey, J., and Spall, H., 1975. Pre-Mesozoic plate tectonics: How far back in Earth history can the Wilson Cycle be extended? Geology, 3, 422-424.

Dickinson, K. A., 1974a. Depositional environments as a guide to uranium exploration in south Texas. Econ. Geol., 69, 149 (Abs).

Dickinson, W. R. (Editor), 1974b. Tectonics and Sedimentation. Soc. Econ. Paleont. Mineral., Tulsa, 204 pp.

Dickinson, W. R., 1974c. Plate tectonics and sedimentation. In: Dickinson (Editor) (q. v.), pp. 1-27.

Dietz, R. S., and Holden, J. C., 1974. Collapsing continental rises: Actualistic concept of geosynclines: A review. In: Dott and Shaver (Editors) (q. v.), pp. 14-25.

Dimitriyev, L. V., 1972. Possible petrological consequences of the rise of mantle beneath mid-ocean ridges. Geochem. Internat., 1972, 168-172.

Dingle, R. V., and Scrutton, R. A., 1974. Continental breakup and the development of post-Paleozoic sedimentary basins around southern Africa. Geol. Soc. Amer. Bull., 85, 1467-1474.

Dixon, C. J., and Pereira, J., 1973. Plate tectonics and mineralization in the Middle East. Econ. Geol., 68, 137.

Dodge, H. W., Jr., and Powell, J. D., 1975. Depositional environments and uranium potential of upper Cretaceous Fox Hills and Lance formations, Crook County, northeastern Wyoming (Abs). Amer. Assoc. Petrol. Geol. Bull., 59, 908.

Dodson, R. G., Needham, R. S., Wilkes, P. G., Page, R. W., Smart, P. G., and Watchman, A. L., 1974. Uranium mineralization in the Rum Jungle - Alligator Rivers Province, Northern Territory, Australia. In: IAEA (q. v.), pp. 551-568.

Doe, B. R., and Smith, D. K. (Editors), 1972. Studies in Mineralogy and Precambrian Geology. Geol. Soc. Amer., Boulder, 348 pp.

Doi, K., Hirono, S., and Sakamaki, Y., 1975. Uranium mineralization by ground water in sedimentary rocks, Japan. Econ. Geol., 70, 628-646.

Dooley, J. R., Jr., Harshman, E. N., and Rosholt, J. N., 1974. Uranium-lead ages of uranium deposits of the Gas Hills and Shirley Basin, Wyoming. Econ. Geol., 69, 527-531.

Dostal, J., Capedri, S., and Aumento, F., 1975. Uranium as an indicator of the origin of Tethyan ophiolites. Earth Planet. Sci. Ltr., 26, 345-352.

Downs, W. F., and Runnells, D. D., 1975. Trace element concentrations in pyrite from sandstone uranium deposits (Abs). Econ. Geol., 70, 1320.

Drake, C. L., 1970. A long-range program of solid Earth studies. Trans. Amer. Geophys. Union, 51, 152-159.

Drake, C. L. (Editor), 1973. U. S. Program for the Geodynamics Project: Scope and Objectives. Nat. Acad. Sci., Washington, 235 pp.

Drozd, R. J., Hohenberg, C. M., and Morgan, C. J., 1974. Heavy rare gases from Rabbit Lake (Canada) and the Oklo Mine (Gabon): Natural spontaneous chain reactions in old uranium deposits. Earth Planet. Sci. Ltr., 23, 28-33.

Dunham, K., 1974. Geochemie erzführender Provinzen in phanerozoischen Platformen. In: Petrascheck (Editor) (q. v.), pp. 29-40.

Eardley, A. J., 1963. Relation of uplifts to thrusts in Rocky Mountains. In: The Backbone of the Americas: Tectonic History from Pole to Pole. Amer. Assoc. Petrol. Geol. Mem. 2.

Eargle, D. H., and Weeks, A. M. D., 1973. Geologic relations among uranium deposits, south Texas, coastal plain region, U. S. A. In: Amstutz and Bernard (Editors) (q. v.), pp. 101-113.

Eargle, D. H., Hinds, G. W., and Weeks, A. M. D., 1971. Uranium gologyy and mines, South Texas. Houston Geol. Soc. Fld. Trp., 59 pp.

Eargle, D. H., Dickinson, K. A., and Davis, B. O., 1975. South Texas uranium deposits. Amer. Assoc. Petrol. Geol. Bull., 59, 766-779.

El-Shazly, E. M., El-Hazek, N. M. T., Abdel-Monem, A. A., Khawasik, S. M., Zayed, Z. M., Nostafa, M. E. M., and Morsi, M. A., 1974. Origin of uranium in Oligocene Qatrani sediments, Western Desert, Arab Republic of Egypt. In: IAEA (q. v.), pp. 467-478.

Embleton, B. J. J., and McElhinny, M. W., 1975. The paleoposition of Madagascar: Paleomagnetic evidence from the Isalo group. Earth Planet. Sci. Ltr., 27, 329-341.

Engel, C. G., and Fisher, R. L., 1975. Granitic to ultramafic rock complexes of the Indian Ocean ridge system, western Indian Ocean. Geol. Soc. Amer. Bull., 86, 1553-1578.

Engel, A. E. J., Itson, S. P., Engel, C. G., Stickney, D. M., and Cray, E. J., Jr., 1974. Crustal evolution and global tectonics:

A petrogenic view. Geol. Soc. Amer. Bull., 85, 843-858.

Ervin, C. P., and McGinnis, L. D., 1975. Reelfoot rift: Reactivated precursor to the Mississippi embayment. Geol. Soc. Amer. Bull., 86, 1287-1295.

Evans, A. M., 1975. Mineralization in geosynclines: The Alpine enigma. Min. Dep., 10, 254-260.

Faure, H., 1972. Paleodynamique du craton africain. Internat. Geol. Cong. Abstr., 24, 75.

Fein, C. D., 1973. Chemical composition and submarine weathering of some oceanic tholeiites from the East Pacific Rise. Geol. Soc. Amer. Abstr. Prog., 5, 41.

Finch, W. I., 1967. Geology of epigenetic uranium deposits in sandstone in the United States. U. S. Geol. Surv. Prof. Ppr. 538, 121 pp.

Finch, W. I., 1975. Uranium in west Texas (Abs). Amer. Assoc. Petrol. Geol.,Bull., 59, 909.

Finch, W. I., Butler, A. P., Jr., Armstrong, F. C., and Weissenborn, A. E., 1973. Uranium. In: Brobst and Pratt (Editors) (q. v.), pp. 456-468.

Fischer, R. P., 1942. Vanadium deposits of Colorado and Utah. U. S. Geol. Surv. Bull. 936-P.

Fischer, R. P., 1950. Uranium-bearing sandstone deposits of the Colorado Plateau. Econ. Geol., 45, 1-11.

Fischer, R. P., 1956. Uranium-vanadium-copper deposits on the Colorado Plateau. U. S. Geol. Surv. Prof. Ppr. 300, 143-154.

Fischer, R. P., 1960. Vanadium-uranium deposits of the Rifle Creek area, Garfield County, Colorado. U. S. Geol. Surv. Bull. 1101, 52 pp.

Fischer, R. P., 1968. The uranium and vanadium deposits of the Colorado Plateau. Ore Deposits of the United States, 1933-1967, 1, 735-746.

Fischer, R. P., 1970. Similarities, differences, and some

genetic problems of the Wyoming and Colorado Plateau types of uranium deposits in sandstone. Econ. Geol., 65, 778-784.

Fischer, R. P., 1974a. Guides to new uranium districts and belts: Exploration targets for tomorrow. Econ. Geol., 69, 149.

Fischer, R. P., 1974b. Exploration guides to new uranium districts and belts. Econ. Geol., 69, 362-376.

Fisher, D. E., 1973. Achondritic uranium. Earth Planet. Sci. Ltr., 20, 151-156.

Francheteau and 7 others, 1975. See ARCYANA.

Frietsch, R., 1974. The iron ores of the Krivoi Rog area, Ukraine Soviet Union. Geol. Foren. Forh., 96, 45-51.

Froidevaux, C., 1973. Energy dissipation and geometric structure at speading plate boundaries. Earth Planet. Sci. Ltr., 20, 419-424.

Froidevaux, C. and Schubert, G., 1975. Plate motion and structure of the continental asthenosphere: A realistic model of the upper mantle. J. Geophys. Res., 80, 2553-2564.

Gabelman, J. W., 1970a. Metallotectonic control of uranium distribution. In: IAEA (q. v.), pp. 187-204.

Gabelman, J. W., 1970b. Speculations on the uranium ore fluid. In: IAEA (q. v.), pp. 315-330.

Gabelman, J. W., 1971. Sedimentology and uranium prospecting. Sed. Geol., 6, 145-186.

Gangloff, A., 1970. Notes sommaires sur la geologie des principaux districts uraniferes etudies par la CEA. In: IAEA (q. v.), pp. 77-87.

Gershoyg, Yu. G., and Kaplun, Ye. Ya., 1970. Occurrences of sulfide mineralization in the Saksagan granites of Krivoy Rog. Geol. Ore Dep., 12(1), 54-62.

Gibb, R. A., 1975. Collision tectonics in the Canadian Shield? Earth Planet. Sci. Ltr., 27, 378-382.

Gill, J. E. (Editor), 1972. Mineral Deposits. 24th Internat.

Geol. Cong., Ottawa, 553 pp.

Gill, J., 1973. Are island arc magmas derived from underthrust lithosphere? Trans. Amer. Geophys. Union, 54, 506.

Gill, R. C. O., and Bridgwater, D., 1975. Geochemistry and significance of early Archean low-K and related tholeiite dykes cutting granitic gneisses in West Greenland. Geol. Soc. Amer. Abstr. Prog., 7, 763.

Gilluly, J., 1971. Plate tectonics and magmatic evolution. Geol. Soc. Amer. Bull., 82, 2383-2396.

Gilluly, J., 1972. Tectonics involved in the evolution of mountain ranges. In: Robertson (Editor) (q. v.), pp. 406-439.

Gingrich, J. E., 1974. Results from a new uranium exploration method. Econ. Geol., 69, 150 (Abs).

Girdler, R. W., 1975. The great negatvie Bouguer gravity anomaly over Africa. Trans. Amer. Geophys. Union, 56, 516-519.

Glaeser, D. J., 1974. Use of fluvial models for predicting sedimentary uranium occurrences (Abs). Econ. Geol., 69, 1180.

Glickson, A. Y., and Lambert, I. B., 1973. Relations in space and time between major Precambrian shield units: An interpretation of Western Australian data. Earth Planet. Sci. Ltr., 20, 395-403.

Goldich, S. S., Ladiak, E. G., Hedge, C. E., and Walthall, F. G., 1966. Geochronology of the midcontinent region, United States: Northern area. J. Geophys. Res., 71, 5389-5408.

Gordon, E., 1974. Uranium: New development is targeted at the future generating marketing. Eng. Min. J., 175(3), 156-160.

Gordon, E., 1975. Uranium: Prices firm up in '74; new foreign policies affect investment. Eng. Min. J., 176(3), 213-217.

Gorlov, N. V., 1975. Geotectonic pattern of the oldest regions of the continental crust: Tentative interpretation of the problem. Proc. USSR Acad. Sci., Geol., 75(2), 13-27.

Granger, H. C., 1966. Ferroselite in a roll-type uranium deposit, Powder River Basin, Wyoming, U. S. Geol. Surv. Prof. Ppr.

550-C, pp. 133-136.

Granger, H. C., 1968. Localization and control of uranium deposits in the southern San Juan mineral belt, New Mexico: An hypothesis. U. S. Geol. Surv. Prof. Ppr. 600-B, pp. 60-70.

Granger, H. C., and Warren, C. G., 1969. Unstable sulfur compounds and the origin of roll-type uranium depoiits. Econ. Geol., 64, 160-163.

Granger, H. C., and Warren, C. G., 1974. Zoning in the altered tongue associated with roll-type uranium deposits. In: IAEA (q. v.), pp. 185-200.

Granger, H. C., Santos, E. S., Dean, B. G., and Moore, F. B., 1961. Sandstone-uranium deposits at Ambrosia Lake, New Mexico: An interim report. Econ. Geol., 56, 1179-1210.

Grauert, B., Seitz, M. G., and Soptrajanova, G., 1974. Uranium and lead gain of detrital zircon studied by isotopic analyses of fission-track mapping. Earth Planet. Sci. Ltr., 21, 389-399.

Green, A. G., 1975. On the postulated Hawaiian plume with emphasis on the limitations of seismic arrays for detecting deep mantle structure. J. Geophys. Res., 80, 4028-4036.

Griffin, V. S., Jr., 1970. Relevancy of the Dewey-Bird hypothesis of cordilleran-type mountain belts and the Wegmann stockwork concept. J. Geophys. Res., 75, 7504-7507.

Griffin, V. S., Jr., 1974. Plate tectonics and the Ouachita system of Texas, Oklahoma, and Arkansas: Discussion. Geol. Soc. Amer. Bull., 85, 145-146.

Griggs, D. T., 1972. The sinking lithosphere and the focal mechanism of deep earthquakes. In: Robertson (Editor) (q. v.), pp. 361-384.

Gruner, J. W., 1956. Concentration of uranium in sediments by multiple migration-accretion. Econ. Geol., 51, 495-520.

Grutt, E. W., 1972. Prospecting criteria for sandstone-type uranium deposits. In: Bowie et al. (Editors) (q. v.), pp. 47-76.

Gvirtzman, G., Friedman, G. M., and Miller, D. S., 1973. Control and distribution of uranium in coral reefs during diagenesis. J. Sed. Petrol., 43, 985-997.

Hackett, J. P., and Bischoff, J. L., 1973. New data on the stratigraphy, extent, and geologic history of the Red Sea geothermal deposits. Econ. Geol., 68, 553-564.

Hafenfeld, S. R., and Brookins, D. G., 1975. Mineralogy of uranium deposits northeast of Laguna district, Sandoval County, New Mexico (Abs). Amer. Assoc. Petrol. Geol. Bull., 59, 910-911.

Haines, E. D., and Zartman, R. E., 1973. Uranium concentration and distribution in six peridotite inclusions of probable mantle origin. Earth Planet. Sci. Ltr., 20, 45-53.

Haji-Vassiliou, A., and Kerr, P. F., 1972. Uranium-organic matter association at La Bajada, New Mexico. Econ. Geol., 67, 41-54.

Haller, J., 1971. Geology of the East Greenland Caledonides. Wiley-Interscience, London, 413,pp.

Hammond, A. L., 1975a. Minerals and plate tectonics: A conceptual revolution. Science, 189, 779-781.

Hammond, A. L., 1975b. Minerals and plate tectonics (II): Seawater and ore formation. Science, 189, 868, 869, 915, 917.

Harrison, J. E., 1972. Precambrian Belt basin of northwestern United States: its geometry, sedimentation, and copper occurrences. Geol. Soc. Amer. Bull., 83, 1215-1240.

Harshman, E. N., 1968. Uranium deposits of the Shirley Basin, Wyoming. Ore Deposits of the United States, 1933-1967, 1, 849-856.

Harshman, E. N., 1972. Uranium ore rolls in the United States. In: IAEA (q. v.), pp. 219-227.

Harshman, E. N., 1974. Distribution of elements in some roll-type uranium deposits. In: IAEA (q. v.), pp. 169-183.

Harshman, E. N., 1974b. Distribution of some elements in roll-type uranium deposits in Texas, Wyoming, and South Dakota. Econ. Geol., 69, 150.

Hayashi, S., 1965. Uranium ore deposits and geology in the Tono area, Gifu Pref. J. Atomic En. Soc. Japan, 7, 74-77.

Hayashi, S., 1970. Uranium occurrences in small sedimentary basins in Japan. In: IAEA (q. v.), pp. 233-239.

Haynes, S. J., and McQuillan, H., 1974. Evolution of the Zagros suture zone, southern Iran. Geol. Soc. Amer. Bull., 85, 739-744.

Heier, K. S., and Rhodes, J. M., 1966. Thorium, uranium, and potassium concentrations in granites and gneisses of the Rum Jungle Complex, N. T., Australia. Econ. Geol., 61, 563-571.

Heinrich, E., 1958. Mineralogy and Geology of Radioactive Raw Material. McGraw-Hill, New York.

Heirtzler, J. R. (Editor), 1972. Understanding the Mid-Atlantic Ridge. National Academy of Science, Washington, 131 pp.

Helmberger, D. V., and Engen, G. R., 1974. Upper mantle shear structure. J. Geophys. Res., 79, 4017-4028.

Henderson, P., MacKinnon, A., and Gale, N. H., 1971. The distribution of uranium in some basic igneous cumulates and its petrological significance. Geochim. Cosmochim. Acta, 35, 917-922.

Herbosch, A., 1974. Facteurs controlant la distribution des elements dans les shales uraniferes de bassin Permien de Lodeve (Herault, France). In: IAEA (q. v.), pp. 359-380.

Hess, F. L., 1914. A hypothesis for the origin of carnotite of Colorado and Utah. Econ. Geol., 9, 675-688.

Hills, E. S., 1972. Time and space relationships of basins of epeirogenic origin in Australia. Internat. Geol. Cong. Abstr., 24, 80.

Hilpert, L. S., 1969. Uranium resources of northwestern New Mexico. U. S. Geol. Surv. Prof. Ppr. 606, 166 pp.

Hoffman, P. F., 1969. Proterozoic paleocurrents and depositional history of the East Arm fold belt, Great Slave Lake. Can. J. Earth Sci., 6, 441-448.

Hoffman, P. F., 1973. Evolution of an early Proterozoic continental margin: The Coronation geosynclines and associated aulacogens of the NW Canadian Shield. Phil Trans. Roy. Soc. Lond., 1-12.

Hostetler, P. B., and Garrels, R. M., 1962. Transportation and precipitation of uranium and vanadium at low temperatures, with special reference to sandstone-type uranium deposits. Econ. Geol., 57, 137-167.

Huff, L. C., and Lesure, F. G., 1962. Diffusion feature of uranium-vanadium deposits in Montezuma Canyon, Utah. Econ. Geol., 57, 226-237.

Hurley, P. M., 1974. Pangeaic orogenic system. Geology, 2, 373-376.

Hurley, P. M., Boudda, A., Kanes, W. H., and Nairn, A. E. M., 1974. A plate tectonics origin for late Precambrian - Paleozoic orogenic belt in Morocco. Geology, 2, 343-344.

Hurst, R. W., Bridgwater, D., Collerson, K. D., and Wetherill, G. W., 1975. 3600-m. y. Rb-Sr ages from very early Archaean gneisses from Saglek Bay, Labrador. Earth Planet. Sci. Ltr., 27, 393-403.

Hutchinson, R. W., and Hodder, R. W., 1972. Possible tectonic and metallogenic relationship between porphyry copper and massive sulfide deposits. Can. Min. Metall. Bull., 65, 34-40.

IAEA, 1970. Uranium Exploration Geology. Internat. Atomic En. Agcy, Vienna.

IAEA, 1971. Peaceful Uses of Atomic Energy. Internat. Atomic En. Agcy., Vienna.

IAEA, 1972. Uranium Exploration Geology. Internat. Atomic En. Agcy., Vienna.

IAEA, 1973. Uranium Exploration Methods. Internat. Atomic En. Agcy., Vienna, 320 pp.

IAEA, 1974. Formation of Uranium Ore Deposits. Internal Atomic En. Agcy., Vienna, 784 pp.

Ingerson, E., 1974. The possibility of geochemical provinces in the ocean basins. In: Petrascheck (Editor) (q. v.), pp. 52-64.

Irving, E., and McGlynn, J. C., 1975. Precambrian paleomagnetism of the northern continents and its tectonic implications. Mtg. Roy. Soc. Lond., Mar. 13-14.

Isaacson, P. E., 1975. Evidence for a western extracontinental land source during the Devonian period in the central Andes. Geol. Soc. Amer. Bull., 86, 39-46.

Isachsen, Y. W., Mitcham, T. W., and Wood, H. B., 1955. Age and sedimentary environments of uranium host rocks, Colorado Plateau. Econ. Geol., 50, 127-134.

Ivankin, P. F., Fotiadi, E. E., and Shchgelov, A. P., 1974. Experimental modeling of the tectonosphere of mobile belts. Geotectonics, 5, 278-287.

Jackson, D., Jr., 1975. Rio Algom's Lisbon uranium mine in Utah opens up new area of ore potential. Eng. Min. J., 176(4), 92-95.

Jahn, B-M, and Murthy, V. R., 1975. Rb-Sr ages of the Archean rocks from the Vermilion district, northeastern Minnesota. Geochim. Cosmochim. Acta, 39, 1679-1689.

Jakes, P., and White, A. J. R., 1972. Major and trace element abundances in volcanic rocks of orogenic areas. Geol. Soc. Amer. Bull., 83, 29-40.

Jayaram, K. M. V., Dwivedy, K. K., Bhurat, M. C., and Kulshrestha, S. G., 1974. A study of the influence of microflora on the genesis of uranium occurrences at Udaisagar, Udaipur District, Rajasthan. Proc. Symp. Form. Uran. Ore Dep., 89-98.

Jenkins, D. A. L., 1974. Detachment tectonics in western Papua New Guinea. Geol. Soc. Amer. Bull., 85, 533-548.

Jensen, M. L., 1958. Sulfur isotopes and the origin of sandstone-type uranium deposits. Econ. Geol., 53, 598-616.

Jolley, W. T., 1975. Subdivision of the Archean lavas of the Abitibi area, Canada, from Fe-Mg-Ni-Cr relations. Earth Planet. Sci.

Ltr., 27, 200-210.

Jordan, T. H., 1974. Some comments on tidal drag as a mechanism for driving plate montions. J. Geophys. Res., 79, 2141-2142.

Jordan, T. H., 1975. The continental tectosphere. Rev. Geophys. Space Phys., 13(3), 1-12.

Junta de Energia Nuclear, 1968. Structural and tectonic synthesis of the central uranium provinces of Portugal. In: Stemprok (Editor) (q. v.), pp. 71-84.

Kamiyama, T., Okada, S., and Shimazaki, Y., 1973. Exploration of uranium deposits in Tertiary conglomerates and snndstones in Japan. In: IAEA (q. v.), pp. 45-55.

Kampschuur, W., and Rondeel, H. E., 1975. The origin of the Betic orogen, southern Spain. Tectonophys., 27, 39-56.

Kaplan, H., Uz, S., and Cetinturk, I., 1974. Le gite d'uranium de Fakili et sa formation. In: IAEA (q. v.), pp. 453-465.

Karig, D. E., 1974. Tectonic erosion at trenches. Earth Planet. Sci. Ltr., 21, 209-212.

Karig, D. E,. and Sharman, G. F., III, 1975. Subduction and accretion in trenches. Geol. Soc. Amer. Bull., 86, 377-389.

Kashirtseva, M. F., 1964. Mineral and geochemical zonation of infiltration uranium ores. Soviet Geol., 10, 51-56.

Kashirtseva, M. F., and Sidel'nikova, V. D., 1973. Distribution of selenium, uranium, and molybdenum during infiltration mineralization. Geol. Ore. Dep., 15(3), 82-92.

Katayama, N., Kubo, K., and Hirono, S., 1974. Genesis of uranium deposits of the Tono mine, Japan. In: IAEA (q. v.), pp. 437-452.

Katz, M. B., 1974. Paired metamorphic belts in Precambrian granulite rocks in Gondwanaland. Geology, 2, 237-241.

Katz, M. B., 1976. Broken Hill: A Precambrian hot spot? Precamb. Res., 3, 91-106.

Kaula, W. M., 1972. Global gravity and tectonics. In:

Robertson (Editor) (q. v.), pp. 386-405.

Kaula, W. M., 1975. Absolute plate motions by boundary velocity minimizations. J. Geophys. Res., 80, 244-248.

Keller, G. R., and Shurbet, D. H., 1975. Crustal structure of the Texas Gulf Coastal Plani. Geol. Soc. Amer. Bull,, 86, 807-810.

Keller, G. R., Smith, R. B., and Braile, L. W., 1975. Crustal structure along the Great Basin - Colorado Plateau transition from seismic refraction studies. J. Geophys. Res., 80, 1093-1098.

Kelley, V. C., 1955. Regional tectonics of the Colorado Plateau and relationship of the origin and distribution of uranium. Univ. New Mex. Publ. Geol., 5.

Kelley, V. C., 1956. Influence of regional structure and tectonic history upon the origin and distribution of uranium of the Colorado Plateau. U. S. Geol. Surv. Prof. Ppr. 300, 171-178.

Kennedy, M. J., 1975. Repetitive orogeny in the northeastern Appalachians: New plate models based upon Newfoundland examples. Tectonophys., 28, 39-87.

Kerr, P. F., 1958. Uranium emplacement in the Colorado Plateau. Geol. Soc. Amer. Bull., 69, 1075-1111.

Key, R. M., and Hutton, S. M., 1976. A tectonic generation of the Limpopo Mobile Belt, and a definition of its western extremity. Precamb. Res., 3, 79-90.

Khain, V. E., 1972. Main trends in the evolution of the Earth's crust (lithosphere). Internat. Geol. Cong. Abstr., 24, 84.

Kidd, D. F., 1942. The silver-pitchblende deposits near Great Bear Lake, N. S. T. In: Newhouse (Editor), (q. v.), pp. 238-239.

Killeen, P. G., and Heier, K. S., 1975. A uranium and thorium enriched province of the Fennoscandian shield in southern Norway. Geochim. Cosmochim. Acta, 39, 1515-1524.

King, P. B., 1975. Ancient southern margin of North America. Geology, 3, 732-734.

Kingma, J. T., 1974. The Geological Structure of New Zealand.

Wiley-Interscience, New York, 407 pp.

Klohn, M. L., and Pickens, W. R., 1974. Geology of the Felder uranium deposit, Live Oak County, Texas. Econ Geol., 69, 151.

Knipping, H. D., 1974. The concepts of supergene versus hypogene emplacement of uranium at Rabbit Lake, Saskatchewan, Canada. In: IAEA (q. v.), pp. 531-549.

Koide, H., and Bhattacharji, S., 1975a. Formation of fractures around magmatic intrusions and their role in ore localization. Econ. Geol., 70, 781-799.

Koide, H., and Bhattacharji, S., 1975b. Mechanistic interpretation of rift valley formation. Science, 189, 791-793.

Komarov, A. N., and Yergorov, V. P., 1970. Radiographic study of uraniferous pyrites. Geol. Ore Dep., 12(4), 109-111.

Komarov, A. N., Zmitkov, A. S., Dmitriev, L. V., and Leonova, L. L., 1973. Uranium distribution in ultrabasics of Indian Ocean rift zones. Geochem. Internat., 10, 217-222.

Konstantynowicz, E., 1972. Mineral distribution in the Permian formations of western Poland. In: Gill (Editor) (q. v.), pp. 373-380.

Kornechuk, E., and Burtek, T., 1974. Lithological features and facies of uranium ore deposits in formations in the Socialist Republic of Romania. In: IAEA (q. v.), pp 343-357.

Korolev, K. G., and Rumyantseva, G. V., 1974. On conditions for the collomorph uranetitanite and brannerite deposition. Proc. USSR Acad. Sci., Geol., 74(9), 76-86.

Korolev, K. G., Rumyantseva, G. V., and Golikova, G. A., 1974. To the problem of uranium transport in hydrothermal emvironments (from experimental data). Proc. USSR Acad. Sci., 74(10), 102-113.

Krauskopf, K. B., 1955. Sedimentary deposits of rare metals. Econ. Geol., 50A, 411-463.

Krauskopf, K. B., 1956. Factors controlling the concentrations of thirteen rare metals in seawater. Geochim. Cosmochim. Acta, 9, 1-32.

Kresten, P.,,1974. Uranium in kimberlites and associated rocks with special reference to Lesotho occurrences. Lithos, 7(3), 171-180.

Kröner, A., and Welin, E., 1974. Evidence for a 500 m. y. old thermal episode in southern South-West Africa. Earth Planet Sci. Ltr., 21, 149-152.

Kropotkin, P. N., 1972. Eurasia as a composite continent. Trans. Amer. Geophys. Union, 53, 180-181.

Krupennikov, V. A., 1969. The distribution features of uranium ore in deposits in carbonaceous-cherty slates and in limestone. Geol. Ore Dep., 11(4), 78-87.

Ladeira, E. A., and Leal, E. D., 1972. Phosphate rock of the Cardo do Abaete region, Minas Gerais State, Brazil. In: Gill (Editor) (q. v.), pp. 435-444.

Lambert, I. B., and Wyllie, P. J., 1970. Low-velocity zone of the Earth's mantle: incipient melting caused by water. Science, 196, 764-766.

Lamey, C. A., 1966. Metallic and Industrial Mineral Deposits. McGraw-Hill, New York, 567 pp.

Lancelot, J. R., Vitrac, A., and Allegre, C. J., 1975. The Oklo natural reactor: Age and evolution studies by U-Pb and Rb-Sr systematics. Earth Planet. Sci. Ltr., 25, 189-196.

Langford, F. F., 1974. A supergene origin for vein-type uranium ores in the light of Western Australia calcrete-carnotite deposits. Econ. Geol., 69, 516-526.

Lapp, R. E., 1975. We may find ourselves short of uranium, too. Fortune, Oct., 1975, pp. 151, 152, 194, 196, 199.

Law, B. E., Barnum, B. E., and Galyardt, G. L., 1975. Tectonic implications of the Fort Union Formation, northwestern Powder River Basin, Wyoming and Montana. Geol. Soc. Amer. Abstr. Prog., 7, 1163.

Lee, M. J., Mukhopadhyay, B., and Brookins, D. G., 1975a. Clay mineralogy of uranium-organic enriched and barren zones in Morrison formation, Ambrosia Lake district, New Mexico (Abs). Amer. Assoc. Petrol. Geol. Bull., 59, 914.

Lee, M. J., Brookins, D. G., and Mukhopadhyay, B., 1975b. Rb-Sr geochronologic study of the Westwater Canyon Member, Morrison Formation (Late Jurassic), Grants Mineral Belt, New Mexico (Abs). Econ. Geol., 70, 1324.

Le Pichon, X., Hyndman, R. D., and Pautot, G., 1971. Geophysical study of the opening of the Labrador Sea. J. Geophys. Res., 76, 4724-4743.

Lesnoi, D. A., and Lepkii, S. D., 1970. In: Pouba and Stemprok (Editors) (q. v.), pp. 299.

Little, H. V., 1974. Uranium deposits in Canada: Their exploration, reserves, and potential. Can. Inst. Min. Metall. Bull., 67, 155-162.

Lloyd, D. C. J., 1975. Uranium. In: J. Spooner, L. Williams, A. Kennedy, and M. Spriggs (Editors), Mining Annual Review. Mining Journal, London, pp. 96-98.

Lowe, D. R., 1975. Regional controls on silica sedimentation in the Ouachita system. Geol. Soc. Amer. Bull., 86, 1123-1127.

Ludwig, K. R., 1975. Uranium-lead isotope apparent ages of pitchblendes, Shirley Basin, Wyoming (Abs). Amer. Assoc. Petrol. Geol. Bull., 59, 1975.

Lukacs, E., Florjancic, A., 1974. Uranium ore deposits in the Permian sediments of northwest Yugoslavia. In: IAEA (q. v.), pp. 313-329.

Lutts, B. G., and Mineyeva, I. G., 1974. Uranium and thorium in Siberian kimberlites. Geochem. Internat., 10, 1278-1281.

Mabile, J., 1968. Long-range trend for uranium Proc. Conf. Nuc. Fuel Explor. Pwr. Rctr. Southern Interstate Nuclear Board, Atlanta.

MacKevett, E. M., Jr., 1963. Geology and ore deposits of the Bokan Mt. uranium-thorium area, southeastern Alaska. U. S. Geol.

Surv. Bull., 1154, 154 pp.

Mackin, J. H., and Schmidt, D. L., 1956. Uranium and thorium bearing minerals in placer deposits in Idaho. U. S. Geol. Surv. Prof. Ppr. 300, pp. 375-380.

Mammerickx, J., Anderson, R. N., Menard, H. W., and Smith, S. M., 1975. Morphology and tectonic evolution of the east-central Pacific. Geol. Soc. Amer. Bull., 86, 111-118.

Markov, C., and Ristic, M., 1974. Characteres mineralogeochimiques et genese du gisement d'uranium de Zirovski Vrh. In: IAEA (q. v.), pp. 331-341.

Marlow, M. S., Scholl, D. W., and Garrison, L. E., 1973. Comparative geologic histories of the Aleutian and Lesser Antilles island arcs. Geol. Soc. Amer. Abstr. Prog., 5, 77-78.

Martin-Calvo, M., 1974. Consideraciones sobre el papel que desempenan las sustancias organicas naturales de caracter humico en la concentracion del uranio. In: IAEA (q. v.), pp. 125-137.

Martin-Delgado-Tamayo, J., and Fernandez-Polo, J. A., 1974. Analogias y diferencias de caracteres de favorabilidad en distintos terrenos sedimentarios de la Cordillera Iberica. In: IAEA (q. v.), pp. 479-491.

Mashchak, M. S., 1969. Occurrences of copper-sulfide mineralization in Proterozoic diabase dikes of the southern slope of the Anabar shield. Geol. Ore Dep., 11(6), 74-78.

Maughan, E. K., 1975. Organic carbon in shale beds of Phosphoria Formation (Abs). Amer. Assoc. Petrol. Geol. Bull., 59, 916-917.

McConnell, R. L., 1975. Biostratigraphy and depositional evvironments of algal stromatolites from the Mescal limestone (Proterozoic) of central Arizona. Precamb. Res., 2, 317-328.

McDougall, I., 1971. Volcanic island chains and sea-floor spreading. Nature, 231, 141-144.

McElhinny, M. W., 1972. Paleomagnetism and plate tectonics.

Internat. Geol. Cong. Abstr., 24, 92.

McGetchin, T. R., Nikhanj, Y. S., and Chodos, A. A., 1973. Carbonate-kimberlite relations in the Cane Valley diatreme, San Juan County, Utah. J. Geophys. Res., 78, 1854-1869.

McKelvey, V. E., and Nelson, J. M., 1950. Characteristics of marine uranium-bearing sedimentary rocks. Econ. Geol., 45, 35-53.

McKelvey, V. E., and Carswell, L. D., 1956. Uranium in the Phosphoria formation. USGS Prof. Ppr. 300, 483-487.

McKelvey, V. E., Everhart, D. L., and Garrels, R. M., 1955. Origin of uranium deposits. Econ. Geol., 50A, 464-533.

McKenzie, D. P., 1972. Plate tectonics. In: Robertson (Editor) (q. v.), pp. 323-360.

McWilliams, M. O., and Dunlop, D. J., 1975. Precambrian paleomagnetism: Magnetizations reset by the Grenville orogeny. Science, 190, 269-272.

Menard, H. W., 1973. Does Mesozoic mantle convection still persist? Earth Planet. Sci. Ltr., 20, 237-241.

Miller, J. L., 1955. Uranium ore controls of the Happy Jack deposit, White Canyon, San Juan County, Utah. Econ. Geol., 50, 156-169.

Mineyev, D. A., Mineyeva, I. G., and Tarkhanova, G. A., 1974. Lanthanids: Their range and spectrum in uranium deposits ores. Proc. USSR Acad. Sci., Geol., 74(9), 76-86.

Mitchell, A. H. G., 1975. Evolution and global tectonics: Petrogenic view: Discussion and reply. Geol. Soc. Amer. Bull., 86, 1487-1488.

Mitchell, W. S., and Aumento, F., 1974. A geochemical comparison of minerals of oceanic continental ultramafic origin. J. Geophys. Res., 79, 5529-5532.

Mittempergher, M., 1970. Characteristics of uranium ore genesis in the Permian and Lower Triassic of the Italian Alps. In: IAEA (q. v.), pp. 253-258.

Mittempergher, M., 1974. Genetic characteristics of uranium deposits associated with Permian sandstones in the Italian Alps. In: IAEA (q. v.), pp. 299-312.

Miyashiro, A., 1972. Metamorphism and related magmatism in plate tectonics. Amer. J. Sci., 272, 656.

Miyashiro, A., 1973a. Metamorphism and Metamorphic Belts. Halsted, New York, 492 pp.

Miyashiro, A., 1973b. Paired and unpaired metamorphic belts. Tectonophys., 17, 241-254.

Miyashiro, A., 1974. Volcanic rock series in island arcs and active continental margins. Amer. J. Sci., 274, 321-355.

Miyashiro, A., 1975. Petrology and plate tectonics. Rev. Geophys. Space Phys., 13(3), 94-101.

Modnikov, I. S., and Lebedev-Zinov'yev, A. A., 1969. Relation between dikes and uranium-molybdenum mineralization in some ore deposits located in volcanic rocks. Geol. Ore. Dep., 11(5), 91-97.

Modnikov, I. S., Tarkhanova, G. A., and Chesnokov, L. V., 1971. The relation between uranium-molybdenum and tin-tungsten-molybdenum hydrothermal mineralization. Geol. Ore Dep., 13(2), 92-97.

Modnikov, I. S., Tyrkin, I. K., Florlov, G. I., and Khomyakov, N. A., 1975. Geological-structural characteristics of a molybdenum deposit, with uranium, which is confined to a buried paleovolcanic apparatus. Proc. USSR Acad. Sci., Geol., 75(3), 66-76.

Moench, R. H., and Schlee, J. S., 1967. Geology and uranium deposits of the Laguna district, New Mexico. U. S. Geol. Surv. Prof. Ppr. 519, 117 pp.

Moghal, M. Y., 1974. Uranium in Siwalik sandstones, Sulaiman Range, Pakistan. In: IAEA (q. v.), pp. 383-403.

Molnar, P., and Tapponnier, P., 1975. Cenozoic tectonics of Asia: Effects of a continental collision. Science, 189, 419-426.

Moorbath, S., O'Nions, R. K., and Pankhurst, R. J., 1975. The evolution of early Precambrian crustal rocks at Isua, West Greenland:

Geochemical and isotopic evidence. Earth Planet. Sci. Ltr., 27, 229-239.

Moore, G. W., 1973. Westward tidal lag as the driving force of plate tectonics. Geology, 1, 99-101.

Morey, G. B., and Sims,,P.,K., 1976. Boundary between two Precambrian W terranes in Minnesota and its geologic significance. Geol. Soc. Amer. Bull., 87, 141-152.

Morse, S. A., 1975. Plagioclase lamellae in hypersthene, Tikkoatokhahh Bay, Labrador. Earth Planet. Sci. Ltr., 26, 331-336.

Morton, R. D., 1974. Sandstone-type uranium deposits in the Proterozoic strata of NW Canada. In: IAEA (q. v.), pp. 255-273.

Morton, R. D., and Sassano, G. P., 1972. Structural studies on the uranium deposit of the Fay Mine, Eldorado, Northwest Saskatchewan. Can. J. Earth Sci., 9, 1368-1375.

Motica, J. E., 1968. Geology and uranium-vanadium deposits in the Uravan mineral belt, southwestern Colorado. Ore Deposits of the United States, 1, 805-813.

Muehlberger, W. R., 1965. Late Paleozoic movement along the Texas lineament. New York Acad. Sci. Trans., II(27), 385-392.

Mukhopadhyay, B., Brookins, D. G., and Bolivar, S. L., 1975. Rb-Sr whole-rock study of the Precambrian rocks of the Pedernal Hills, New Mexico. Earth Planet. Sci. Ltr., 27, 283-286.

Murray, C. G., 1970. Magma genesis and heat flow: Differences between mid-oceanic ridges and African rift salleys. Earth Planet. Sci. Ltr., 9, 34-38.

Newhouse, W. H. (Editor), 1942. Ore Deposits as Related to Structural Features. Princeton Univ. Press, Princeton 280 pp. (Hafner, New York, 1969, reprint).

Nininger, R. D., 1974. The world uranium supply challenge: An appraisal. In: IAEA (q. v.), pp. 3-18.

Nitsan, U., 1973. Viscous heat production in a slab. J. Geophys. Res., 78, 1395-1395.

Nitu, G., 1974. Les conditions tectono-magmatiques de la formation des gisements d'uranium de Roumaine. In: IAEA (q. v.), pp. 679-692.

Nkomo, I. T., and Stuckless, J. S., 1975. Uranium-lead systematics and uranium leaching in core samples from Granite Mountains, Wyoming (Abs). Amer. Assoc. Petrol. Geol. Bull., 59, 918.

Noble, D. C., and 6 others, 1975. Chemical and isotopic constraints on the origin of low-silica latite and andesite from the Andes of central Peru. Geology, 3, 501-504.

Oldenburg, D. W., and Brune, J. N., 1975. An explanation for the orthogonality of ocean ridges and transform faults. J. Geophys. Res., 80, 2575-2585.

Omsted, R. W., and Al-Shaieb, Z., 1975. Geochemical anomalies, uranium potential of south-central Oklahoma (Abs). Econ. Geol., 70, 1326.

Omel'yanenko, B. I., Rossman, G. I., Zheleznova, E. I., Elisyeva, O. P., Reznikova, A. M., Skorospelikin, S. A., and Smirnova, A. I., 1973. Behavior of uranium during wallrock alteration. Geol. Ore Dep., 15(5), 60-65.

Oparysheva, L. G., Shmariovich, E. M., Larkin, E. D., and Shchetochkin, V. N., 1973. Peculiarities of the uranium mineralization within the rocks of sedimentary cover and basement granites. Proc. USSR Acad. Sci., Geol., 73(7), 32-43.

Ostle, D., 1970. Criteria for the selection of exploration areas in the United Kingdom. In: IAEA (q. v.), pp. 345-353.

Oversby, V. M., 1975a. Lead isotopic systematics and ages of Archean acid intrusives in the Kalgoorlie-Norseman area, Western Australia. Geochim. Cosmochim. Acta, 39, 1107-1125.

Oversby, V. M., 1975b. Lead isotopic study of aplites from the Precambrian basement rocks near Ibadan, southwestern Nigeria. Earth Planet. Sci. Ltr., 27, 177-180.

Oxburgh, E. R., and Turcotte, D. L., 1974. Membrane tectonics and the East African Rift. Earth Planet. Sci. Ltr., 23, 133-140.

Oyarzun, M. J., 1975. Porphyry copper and tin-bearing porphyries: A discussion of genetic models. Phys. Earth Planet. Int., 9, 259-263.

Pankhurst, R. J., Moorbath, S., McGregor, V. R., 1973a. Late event in the geological evaluation of the Godthaab district, West Greenland. Nature Phys. Sci., 243, 24-27.

Pankhurst, R. J., Moorbath, S., Rex, D. C., and Turner, G., 1973b. Mineral age patterns in ca. 3700-m. y. rocks from West Greenland. Earth Planet. Sci. Ltr., 20, 157-162.

Parak, T., 1973. Rare earths in the apatite iron ore of Lappland together with some data about the Sr, Th, and U content of these ores. Econ. Geol., 68, 210-221.

Park, C. F., Jr., and MacDiarmid, R. A., 1964. Ore Deposits. Freeman, San Francisco, 475 pp.

Perelman, A. I., 1967. Geochemistry of Epigenesis. Plenum, New York, 213 pp.

Perets, N. A., and Bylinskaya, L. V., 1969. Some features of the uranium-zirconium paragenetic association. Geol. Ore Dep., 11(4), 59-66.

Petrascheck, W. E. (Editor), 1974. Metallogenetic and Geochemical Provinces. Springer, Vienna, 183 pp.

Petrascheck, W. E., Erkan, E., and Neuwirth, K., 1974. Permo-Triassic uranium ore in the Austrian Alps: Paleogeographic control as a guide for prospecting. In: IAEA (q. v.), pp. 291-298.

Pickens, W. R., 1974. Uranium in south Texas. Econ. Geol. 69, 151-152 (Abs).

Pidgeon, R. T., and Hopgood, A. M., 1975. Geochronology of Archaean gneisses and tonalites from north of Frederikshåb isblink, S. W. Greenland. Geochim. Cosmochim. Acta, 39, 1333-1346.

Pitcher, W. S., and Berger, A. R., 1972. The Geology of Donegal: A Study of Granite Emplacements and Unroofing. Wiley-Interscience, New York, 435 pp.

Plushkal, O., 1970. Uranium mineralization in the Bohemian massif. In: IAEA (q. v.), pp. 107-116.

Poty, B. P., Leroy, J., and Cuney, M., 1974. Les inclusions fluides dans les minerais des gisements d'uranium intragranitiques du Limousin et du Forez (Massif Central, France). In: IAEA (q. v.), pp. 569-582.

Pouba, Z., and Stemprok, M. (Editors), 1970. Problems of Hydrothermal Ore Deposition. Schweitzerbart'sche Verlag., Stuttgart, 396 pp.

Pretorius, D. A., 1974. The Nature of the Witwatersrand Gold-Uranium Deposits. Inf. Circ. 86, Econ. Geol. Res Unit, Univ. of Witwatersrand, Johannesburg, 50 pp.

Putnam, G. W., 1975. Base metal distribution in granitic rocks: Three-dimensional variation in the Light Creek stock, California. Econ. Geol., 70, 1225-1241.

Radkevich, E. A., 1972. Metallogenic zonality of the Pacific ore belt. Internat. Geol. Cong. Abstr., 24, 144.

Radusinovic, D., 1974. Zletovska Reka uranium deposit. In: IAEA (q. v.), pp. 593-601

Rawson, R. R., 1975. The sabhka environment: A new frontier for uranium exploration (Abs). Econ. Geol., 70, 1327.

Richter, F. M., and Parsons, B., 1975. On the interaction of two scales of convection in the mantle. J. Geophys. Res., 80, 2529-2541.

Robertson, D. S., 1970. Uranium: Its geological occurrence as a guide to exploration. In: IAEA(q. v.), pp. 267-277.

Robertson, E. C. (Editor), 1972. The Nature of the Solid Earth. McGraw-Hill, New York, 677 pp.

Robertson, D. S., 1974. Basal Proterozoic units as fossil time markers and their use in uranium prospection. In: IAEA (q. v.), pp. 495-512.

Robinson, J. L., 1974. A note on convection in the Earth's

mantle. Earth Planet. Sci. Ltr., 21, 190-193.

Robinson, S. C., and Hewitt, D. F., 1958. Uranium deposits of Bancroft region, Ontario. Proc. Internat. Conf. Pfl. Uses Atom. En., 2, 498-501.

Robinson, B. W., and Morton, R. D., 1972. Geology and geochronology of the Echo Bay area, N. W. T. Can. J. Earth Sci., 9, 158-162.

Robinson, B. W., and Ohmoto, H., 1973. Mineralogy, fluid inclusions, and stable isotopes of the Echo Bay U-Ni-Ag-Cu deposits, Northwest Territories, Canada. Econ. Geol., 73, 635-656.

Roddick, J. C., Compston, W., and Durney, D. W., 1976. The radiometric age of the Mount Keith granodiorite, a maximum age estimate for an Archaean greenstone sequence in the Yilgarn block, Western Australia. Precamb. Res., 3, 55-78.

Rogers, J. J. W., and McKay, S. M., 1972. Chemical evolution of geosynclinal material. In: Doe and Smith (Editors) (q. v.), pp. 3-28.

Rogova, V. P., Belova, L. N., Kiziyarov, G. P., and Kuznetsova, N. N., 1974. Calciouranoite, a new hydroxide of uranium. Internat. Geol. Rev., 16, 1255-1256.

Roper, P. J., 1974. Plate tectonics: A plastic as opposed to a rigid body model. Geology, 2, 247-250.

Rose, A. W., 1974. Chloride complexing of copper and silver in the origin of red bed copper, sandstone-type uranium, and related ore deposits. Econ. Geol., 69, 1186 (Abs).

Rosendahl, B. R., Moberly, R., Halunen, A. J., Rose, J. C., and Kroenke, L. W., 1975. Geological and geophysical studies of the Canton Trough region. J. Geophys. Res., 80, 2565-2574.

Ross, D. A., 1972. Red Sea hot brine area revisited. Science, 175, 1455.

Ross, R. C., 1975. Uranium recovery from phosphoric acid nears reality as a commercial uranium source. Eng. Min. J., 176(12), 80-85.

Routhier, P., Brouder, P., Fleischer, R., Macquar, J. C., Pavillon, M. J., Roger, G., and Rouvier, H., 1973. Some major concepts of metallogeny. Min. Dep., 8, 237-258.

Roy, R. F., Blackwell, D. D., and Decker, E. R., 1972. Continental heat flow. In: Robertson (Editor) (q. v.), pp. 506-543.

Ruzicka, V., 1971. Geological comparison between east European and Canadian uranium deposits. Geol. Surv. Can. Ppr. 70-48.

Saggerson, E. P., and Turner, L. M., 1976. A review of the distribution of metamorphism in the ancient Rhodesian craton. Precamb. Res., 3, 1-53.

Samana, J. C., 1973. Ore deposits and continental weathering: A contribution to the problem of geochemical inheritance of heavy metal contents of basement areas and of sedimentary basins. In: Amstutz and Bernard (Editors) (q. v.), pp. 247-265.

Sandusky, C. L., 1975. Sedimentology of Morrison Formation in southern San Juan basin (Abs). Amer. Assoc. Petrol. Geol. Bull., 59, 922.

Santos, G., Jr., 1974. Mineral distribution and geological features of the Philippines. In: Petrascheck (Editor) (q. v.), pp. 89-105.

Sarkar, S. N., 1972. Present status of Precambrian geochronology of peninsular India. 25th Internat. Geol. Cong., 1, 260-268.

Sassano, G. P., Fritz, P., and Morton, R. D., 1972. Paragenesis and isotopic composition of some gangue minerals from the uranium deposits of Eldorado, Saskatchewan. Can. J. Earth Sci., 9, 141-146.

Saucier, A. E., 1975. Paleotectonic setting of late Jurassic Morrison Formation on Colorado Plateau (Abs). Amer. Assoc. Petrol. Geol. Bull., 59, 922.

Scheidegger, A. E., 1972. The rheology of the tectonosphere. Internat. Geol. Cong. Abstr., 24, 99.

Schidlowski, M., Eichmann, R., and Junge, C. E., 1975. Precambrian sedimentary carbonates: Carbon and oxygen isotope geochemistry and implications for the terrestrial oxygen budget. Precamb. Res., 2, 1-69.

Schilling, J. G., and Bonatti, E., 1975. East Pacific Ridge ($2^{o}S$ - $19^{o}S$) versus Nazca interplate volcanism: Rare-earth evidence. Earth Planet. Sci. Ltr., 25, 93-102.

Schluger, P. R., and Roberson, H. E., 1975. Mineralogy and chemistry of the Patapsco formation, Maryland, related to the ground-water geochemistry and flow system: A contribution to the origin of red beds. Geol. Soc. Amer. Bull., 86, 153-158.

Schwab, F. L., 1975. Framework mineralogy and chemical composition of continental margin-type sandstone. Geology, 3, 487-490.

Schwarzer, R. R., and Rogers, J. W., 1973. Alkali olivine basalt series of the Davis Mountains, South Central Texas: A comparison with worldwide occurrences. Geol. Soc. Amer. Abstr. Prog., 5, 102.

Scrutton, R. A., 1973. Structure and evolution of the sea floor south of South Africa. Earth Planet. Sci. Ltr., 19, 250-256.

Seeland, D. A., 1975. Uranium and hydrocarbon exploration target areas suggested by Eocene stream patterns in the Wind River Basin, Wyoming (Abs). Econ. Geol., 70, 1329.

Seitz, M. G., and Hart, S. R., 1973. Uranium and boron distribution in some oceanic ultramafic rocks. Earth Planet. Sci. Ltr., 21, 97-107.

Sharp, W. E., 1974. A plate tectonic origin for diamond-bearing kimberlites. Earth Planet. Sci., Ltr., 21, 351-354.

Sharp, W. N., McKay, E. J., McKeown, F. A., and White, A. M., 1954. Geology and uranium deposits of Pumpkin Butter area of the Powder River Basin, Wyoming. U. S. Geol. Surv. Bull. 1107-H, pp. 541-552.

Shatkova, L. N., and Shatkov, G. A., 1973. Possible source of ore in uranium-fluorite deposits. Geol. Ore Dep., 15(4), 36-43.

Shaw, D. M., Dostal, J., and Keays, R. R., 1976. Additional

estimates of continental surface Precambrian shield composition in Canada. Geochim. Cosmochim. Acta, 40, 73-83.

Shawe, D. R., 1974. Alteration of red beds and deposition of uranium-vanadium ores at Slick Rock, Colorado, by pore water expelled from Upper Cretaceous Mancos shale. Econ. Geol., 69, 152 (Abs).

Shawe, D. R., and Granger, H. C., 1965. Uranium ore rolls: An analysis. Econ. Geol., 60, 240-250.

Shawe, D. R., Hite, R. J., and Inthuputi, B., 1975. Potential for sandstone-type uranium deposits in Jurassic rocks, Khorat Plateau, Thailand. Econ. Geol., 70, 538-541.

Sheppard, D. S., Adams, C. J., and Bird, G. W., 1975. Age of metamorphism and uplift in the Alpine schist belt, New Zealand. Geol. Soc. Amer. Bull., 86, 1147-1153.

Sherman, J. T., 1969. Uranium. Eng. Min. J., 170(3), 104-108.

Sherman, J. T., 1970. Uranium. Eng. Min. J., 171(3), 92-96.

Sherman, J. T., 1971. Uranium. Eng. Min. J., 172(3), 108-111.

Sherman, J. T., 1972, Uranium. Eng. Min. J., 173(3), 140-143.

Shibata, K., and Adachi, M., 1974. Rb-Sr whole-rock ages of Precambrian metamorphic rocks in the Kamiaso conglomerate from central Japan. Earth Planet. Sci. Ltr., 21, 277-287.

Shoemaker, E. M., 1956. Structural features of the central Colorado Plateau and their relation to uranium deposits. U. S. Geol. Surv. Prof. Ppr. 300, 155-170.

Shtreys, N. A., and Tseysler, V. M., 1972. Formation of geosynclinal regions. Geotectonics, 1, 1-2.

Shurbet, D. H., and Cebull, S. E., 1975. The age of the crust beneath the Gulf of Mexico. Tectonophys., 28, T25-T30.

Siegl, W., 1972. Die uranparagenese von Mittelberg (Salzburg, Österreich). Tschemaks. Mineral. Petrog. Mitt., 17, 263-265.

Sighinolfi, G. P., and Sakai, T., 1974. Uranium and thorium in potash-rich rhyolites from western Bahia (Brazil). Chem. Geol., 14, 23-28.

Sillitoe, R. H., 1972a. A plate tectonic model for the origin of porphyry copper deposits. Econ. Geol., 67, 184-197.

Sillitoe, R. H., 1972b. Relation of metal provinces in western America to subduction of oceanic lithosphere. Geol. Soc. Amer. Bull., 83, 813-818.

Sillitoe, R. H., 1972c. Formation of certain massive sulphide deposits at sites of sea-floor spreading. Trans. Inst. Min. Metall., 81, B141-B148.

Sillitoe, R. H., 1973. Geology of the Los Pelambres porphyry copper deposit, Chile. Econ. Geol., 68, 1-10.

Sleep, N. H., 1975. Formation of oceanic crust: Some thermal constraints. J. Geophys. Res., 80, 4037-4042.

Smirnov, V. I., and Tugarinov, A. I., 1969. The uranium deposits of France. Geol. Ore Dep., 11(6), 3-13.

Smith, E. E. N., 1974. Review of current concepts regarding vein deposits or uranium. In: IAEA (q. v.), pp. 515-529.

Smith, R. B., 1975. Significance of calcium carbonate cementation for uranium exploration.(Abs). Amer. Assoc. Petrol. Geol. Bull., 59, 923-924.

Smith, D. K., and Stohl, F. V., 1972. Crystal structure of beta-uranophane. In: Doe and Smith (Editors) (q. v.), pp. 281-288.

Smithson, S. B., and Decker, E. R., 1973. K, U, and Th distribution between dry and wet facies of syenitic intrusion and the role of fluid content. Earth Planet. Sci. Ltr., 19, 131-134.

Spall, H., 1972. Paleomagnetism and Precambrian continental drift. Internat. Geol. Cong. Abstr., 24, 101.

Spencer, A. M. (Editor), 1974. Mesozoic-Cenozoic Orogenic Belts: Data for Orogenic Studies. Scottish Academic Press, Edinburgh, 809 pp.

Spencer, E. W., and Kozak, S. J., 1975. Precambrian evolution of the Spanish Peaks Area, Montana. Geol. Soc. Amer. Bull., 86, 785-792.

Squyres, J. B., 1974. Uranium deposits in the South San Juan Basin, New Mexico. Econ. Geol., 69, 152 (Abs).

Steacy, H. R., Plant, A. G., and Boyle, R. ., 1974. Brannerite associated with native gold at the Richardson mine, Ontario. Can. Mineralog., 12, 360-364.

Steiner, M. B., and Helsley, C. E., 1975. Reversal patterns and apparent polar wander for the late Jurassic. Geol. Soc. Amer. Bull., 86, 1537-1543.

Steinhorn, T. L., 1973. Depth-of-origin of magmas of the high cascades as inferred from geochemical relations. Geol. Soc. Amer. Abstr. Prog., 5, 110.

Stemprok, M. (Editor), 1968. Endogenous Ore Deposits. Academia, Prague, 425 pp.

Stern, C. R., 1974. Melting products of olivine tholeiite basalt in subduction zones. Geology, 2, 227-230.

Stevens, C. H., and Ridley, A. P., 1974. Middle Paleozoic offshelf deposits in southeastern California: Evidence for proximity of the Antler orogenic belt? Geol. Soc. Amer. Bull., 85, 27-32.

Stewart, J. H., and Poole, F. G., 1975. Extension of the Cordilleran miogeosynclinal belt to the San Andreas fault, southern California. Geol. Soc. Amer. Bull., 86, 202-212.

Stewart, J. H., 1976. Late Precambrian evolution of North America: Plate tectonics implications. Geology, 4, 11-15.

Stipanicic, P. N., 1970. Conceptos geostructurales generales sobre la distribucion de los yacimientos uraniferos con control sedimentario en la Argentina y posible aplicacion de los mismos en el resto de Sudamerica. In: IAEA (q. v.), pp. 205-216.

Stockwell, C. H., 1968. Geochronology of stratified rocks of the Canadian Shield. Can. J. Earth Sci., 5, 693-698.

Strand, T., and Kulling, O., 1972. Scandinavian Caledonides. Wiley-Interscience, London, 320 pp.

Strong, D. F., 1973. Volcanic, intrusive, and tectonic activity in an Ordovician island arc of central Newfoundland. Geol. Soc. Amer. Abstr. Prog., 5, 224.

Stuckless, J. S., Bunker, C. M., Bush, C. A., Doering, W. P., and Scott, J. H., 1975. Preliminary geochemical and petrologic studies of core samples from Granite Mountains, Wyoming (Abs). Amer. Assoc. Petrol. Geol.,Bull., 59, 924.

Subbotin, S. I., Sollogub, V. B., Khain, V. Y., Slavin, V. I., and Chekunov, A. V., 1972. Some remarks rencerning the structure and evolution of the Earth's crust. Hung. Geophys. Inst. Geophys. Trans., 1972, 151-165.

Sun, S-S., Tatsumoto, M., Schilling, J-G., 1975. Mantle plume mixing along the Reykjanes Ridge axis: Lead isotopic evidence. Science, 190, 143-147.

Sutton, W. R., and Soonawala, N. M., 1975. A soil radium method for uranium prospecting. Can. Min. Metall. Bull., 68, 51-56.

Svenke, E., 1956. The occurrence of uranium and thorium in Sweden. Proc. Internat. Conf. Pfl. Uses Atom. En., 6, 198-199.

Swanson, V. E., 1961. Geology and geochemistry of uranium in marine black shales: A review. U. S. Geol. Surv. Prof. Ppr. 356-C, 67-112.

Sychev, I. V., Kozyrev, V. N., Modnikov, I. S., Rossman, G. I., and Shvortsova, K. V., 1974. Features of transformation of uranium-molybdenum ores under supergene conditions. Internat. Geol. Rev., 16, 1315-1321.

Symons, D. T. A., 1975. Hudsonian glaciation and polar wander from the Gowganda Formation, Ontario. Geology, 3, 303-306.

Szadeczky-Kardoss, E., 1974. Metallogenesis and distribution of elements around the zones of subduction. In: Petrascheck (Editor) (q. v.), pp. 68-88.

Tatsch, J. H., 1959. Initial results of preliminary investigations (1938-1958) regarding the feasibility of using polyistic planetary configurations for making certain solar-system evolutionary studies. U. S. Embassy, Caracas, Venezuela, in manus.

Tatsch, J. H., 1960. The evolution of the solar system in accordance with a dual primeval planet model. U. S. Embassy, Caracas, Venezuela, private publication, given limited distribution to approximately 300 selected Earth and planetary scientists in various parts of the world.

Tatsch, J. H., 1962. The evolution of the retrograde motion of certain planetary satellites in accordance with a dual primeval planet model, outline of doctoral dissertation, submitted to graduate advisor (the late Gerard P. Kuiper), February 4, 1962, about 75 pages.

Tatsch, J. H., 1963a. Certain selenophysical implications of applying a dual primeval planet model to the Earth. Trans. Amer. Geophys. Union, 44, 877.

Tatsch, J. H., 1963b. Certain seismological implications of applying a dual primeval planet model to the Earth. Trans. Amer. Geophys. Union, 44, 887.

Tatsch, J. H., 1963c. Certain volcanological implications of applying a dual primeval planet model to the Earth. Trans. Amer. Geophys. Union, 44, 892.

Tatsch, J. H., 1964a. Distribution of active volcanoes: Summary of preliminary results of three-dimensional least-squares analysis. Geol. Soc. Amer. Bull., 75, 751-752.

Tatsch, J. H., 1974b. Certain geomagnetic implications of applying a dual primeval planet model to the Earth. Trans. Amer. Geophys. Union, 45, 39.

Tatsch, J. H., 1964c. Certain oceanographical implications of applying a dual primeval planet model to the Earth. Trans. Amer. Geophys. Union, 45, 74.

Tatsch, J. H., 1964d. Further volcanological implications of applying a dual primeval planet model to the Earth. Trans. Amer. Geophys. Union, 45, 124.

Tatsch, J. H., 1965a. The global distribution patterns of fossil volcanoes and their relationship to the global distribution patterns of active volcanoes. National Science Foundation proposal number P51073R.

Tatsch, J. H., 1966a. Certain correlations between geomagnetic observations in the Indian Ocean and deductions arrived at from applying a dual primeval planet model to that region of the Earth. Trans. Amer. Geophys. Union, 47, 56-57.

Tatsch, J. H., 1966b. Certain correlations between oceanographic observations along the South Atlantic Ridge and deductions arrived at from applying a dual primeval planet model to that region of the Earth. Trans. Amer. Geophys. Union, 47, 123.

Tatsch, J. H., 1966c. Certain correlations between geomagnetic observations in the South Atlantic Ocean and deductions arrived at from applying a dual primeval planet model to that region of the Earth. Trans. Amer. Geophys. Union, 47, 464.

Tatsch, J. H., 1966d. Certain correlations between oceanographic observations along the North Atlantic Ridge and deductions made from applying a dual primeval planet model to that region of the Earth. Trans. Amer. Geophys. Union, 47, 477.

Tatsch, J. H., 1966e. Certain correlations between selenophysical observations and deductions arrived at from applying a dual primeval planet model to the Earth-Moon system. Trans. Amer. Geophys. Union, 47, 486.

Tatsch, J. H., 1966f. Certain correlations between seismological observations in the African continent and deductions made from applying a dual primeval planet model to that region of the Earth. Trans. Amer. Geophys. Union, 47, 490.

Tatsch, J. H., 1967. Global geomagnetic evidence supporting the existence of a geophysical equator predicted as a conseqnence of applying a dual primeval planet model to the Earth. Trans. Amer. Geophys. Union, 48, 59.

Tatsch, J. H., 1969. Sea-floor spreading, continental drift, and plate tectonics unified into a single global concept by the application of a dual primeval planet hypothesis to the Earth. Trans. Amer. Geophys. Union, 50, 672.

Tatsch, J. H., 1970. Global seismicity patterns as interpreted in accordance with a dual primeval planet hypothesis. Geol. Soc. Amer. Abstr. Prog., 2, 153.

Tatsch, J. H., 1972a. The Earth's Tectonosphere: Its Past Development and Present Behavior. Tatsch Assoc., Sudbury, 889 pp.

Tatsch, J. H., 1972b. Global symmetry of mineralization belts. Geol. Soc. Amer. Abstr. Prog., 4, 729-730.

Tatsch, J. H., 1973a. Mineral Deposits: Origin, Evolution, and Present Characteristics. Tatsch Assoc., Sudbury, 264 pp.

Tatsch, J. H., 1973b. The origin, evolution, and present characteristics of the Tertiary petroleum deposits of the world. Tatsch Assoc., Sudbury, 58 pp.

Tatsch, J. H., 1973c. The origin, evolution, and present characteristics of Mesozoic petroleum deposits of the world. Tatsch Assoc., Sudbury, 62 pp.

Tatsch, J. H., 1973d. The origin, evolution, and present characteristics of the Paleozoic petroleum deposits of the world. Tatsch Assoc., Sudbury, 63 pp.

Tatsch, J. H., 1973e. The origin, evolution, and present characteristics of the giant oil fields of North America. Tatsch Assoc., Sudbury, 67 pp.

Tatsch, J. H., 1973f. The origin, evolution, and present characteristics of the petroleum deposits in the Arctic Ocean and North Sea areas. Tatsch Assoc., Sudbury, 62 pp.

Tatsch, J. H., 1973g. The origin, evolution, and present characterisitics of the petroleum deposits in the Middle East areas. Tatsch Assoc., Sudbury, 62 pp.

Tatsch, J. H., 1973h. The origin, evolution, and present

characteristics of the petroleum deposits within the Indonesian islands and within other areas of the arc extending roughly from the Bay of Bengal to the Kermadec-Tonga region. Tatsch Assoc., Sudbury, 66 pp.

Tatsch, J. H., 1966i. The origin, evolution, and present characteristics of the Tertiary basins that appear to be largely nonpetroliferous. Tatsch Assoc., Sudbury, 62 pp.

Tatsch, J. H., 1973j. The probability for the existence of Tertiary petroleum deposits between Cape Cod and Cape Fear. Tatsch Assoc., Sudbury, 65 pp.

Tatsch, J. H., 1973k. Petroleum in the Philippines: An analysis of related Mesozoic and Cenozoic geophenomena. Tatsch Assoc., Sudbury, 43 pp.

Tatsch, J. H., 1974a. The Moon: Its Past Development and Present Behavior. Tatsch Assoc., Sudbury, 338 pp.

Tatsch, J. H., 1974b. Petroleum Deposits: Origin, Evolution, and Present Characteristics. Tatsch Assoc., Sudbury, 378 pp.

Tatsch, J. H., 1973c. Reversals of the Earth's magnetic field as interpreted by the Tectonospheric Earth Model. Nagata Conference, June 3-4, Pittsburgh, Penna.

Tatsch, J. H., 1975a. Copper Deposits: Origin, Evolution, and Present Characteristics. Tatsch Assoc., Sudbury, 339 pp.

Tatsch, J. H., 1975b. The Physics of the Solar System: An Analysis Based on the Dual Primeval Planet Hypothesis. Tatsch Assoc., Sudbury, about 450 pp,, in prep.

Tatsch, J. H., 1975c. The Principle of Universal Heterogeneity: The Basis for the Dual Primeval Planet Hypothesis. Tatsch Assoc., Sudbury, about 475 pp., in prep.

Tatsch, J. H., 1975d. The Earth's Tectonosphere: Its Past Development and Present Behavior, 2nd ed. Tatsch Assoc., Sudbury, about 550 pp., in press.

Tatsch, J. H., 1975e. Geothermal Deposits: Origin, Evolution, and Present Characteristics. Tatsch Assoc., Sudbury, about 340 pp., in press.

Tatsch, J. H., 1975f. Gold Deposits: Origin, Evolution, and Present Characteristics. Tatsch Assoc., Sudbury, 279 pp.

Tatsch, J. H., 1975g. Iron Deposits: Origin, Evolution, and Present Characteristics. Tatsch Assoc., Sudbury, about 320 pp., in prep.

Tatsch, J. H., 1975h. The geodynamics of continental interiors as interpreted by a Tectonospheric Earth Model. Penrose Conference, December 14-19, San Diego.

Tatsch, J. H., and Talbot, J. L., 1971. Global symmetry of tectonic belts. Geol. Soc. Amer. Abstr. Prog., 3, 728-729.

Taylor, H. P., Jr., 1975. Stable isotope geochemistry. Rev. Geophys. Space Phys., 13, 102-107.

Thiel, K., Herr, W., and Becker, J., 1972. Uranium distribution in basalt fragments of five lunar samples. Earth Planet. Sci. Ltr., 16, 31-44.

Thomas, W. A., 1973. Southwestern Appalachian structural system beneath the Gulf Coastal Plain. Amer. J. Sci., 273A, 372-390.

Thomas, W. A., 1975. Appalachian-Ouachita structure and plate tectonics (Abs). Geol. Soc. Amer. Abstr. Prog., 7, 543-544.

Thorpe, R. I., 1971. Lead isotope evidence on the age of mineralization, Great Bear Lake. Geol. Surv. Can. Ppr. 71-1B, pp. 72-81.

Timofeyev, A. A., and Tashchilkin, A. A., 1975. Types of marginal structures generated at various stages of geosynclinal evolution. Proc. USSR Acad. Sci., Geol., 75(1), 31-42.

Titayeva, N. A., Filonov, V. A., Ovchenkov, V. Ya., Veksler, T. I., Orova, A. V., and Tyrina, A. S., 1974. Behavior of uranium and thorium isotopes of crystalline rocks and surface waters in a cold wet climate. Geochem. Internat., 10, 1136-1145.

Trofimov, V. S., 1972. On the origin of diamondiferous eclogite. Internat. Geol. Cong. Abstr., 24, 153.

Truswell, J. F., and Eriksson, K. A., 1975. A paleoenvironmental interpretations of the early Proterozoic Malmani dolomite from Zwartkops, South Africa. Precamb. Res., 2, 277-303.

Turcotte, D. L., and Oxburgh, E. R., 1973. Mid-plate tectonics. Nature, 244, 337-339.

Udas, G. R., and Mahadevan, T. M., 1974. Controls and genesis of uranium mineralization in some geological environments in India. In: IAEA (q. v.), pp. 425-436.

Van Breemen, O., Aftalion, M., and Allaart, J. H., 1974. Isotopic and geochronologic studies on granites from the Ketilidian mobile belt of south Greenland. Geol. Soc. Amer. Bull., 85, 403-412.

Van den Linden, W. J. M., 1975a. Crustal attenuation and sea-floor spreading in the Labrador Sea. Earth Planet. Sci. Ltr., 27, 409-423.

Van den Linden, W. J. M., 1975b. Mesozoic and Cainozoic opening of the Labrador Sea, the North Atlantic, and the Bay of Biscay. Nature, 253, 320-324.

Van Houten, F. B., 1973. Origin of red beds: A review. In: F. A. Donath et al. (Editors), Annual Review of Earth and Planetary Sciences. Ann. Rev., Palo Alto, pp. 39-61.

Veevers, J. J., Powell, C. M., and Johnson, B. D., 1975. Greater India's place in Gondwanaland and in Asia. Earth Planet. Sci. Ltr., 27, 383-387.

Vine, J. D., and Tourtelot, E. B., 1970. Geochemistry of black shale deposits: A summary report. Econ. Geol., 65, 253-258.

Vinnik, L. P., and Lenartovich, E., 1975. Horizontal inhomogeneities in the upper mantle of the Carpathians and Caucasus. Tectonophys., 28, 275-291.

Vogt, P. R., 1975. Changes in geomagnetic reversal frequency at times of tectonic change: Evidence for coupling between core and upper mantle processes. Earth Planet. Sci. Ltr., 25, 313-321.

Von Backström, J. W., 1974a. Uranium deposits in the Karoo Supergroup near Beaufort West, Cape Province, South Africa. In: IAEA (q. v.), pp. 419-424.

Von Backström, J. W., 1974b. Other uranium deposits. In: IAEA (q. v.), pp. 605-624.

Walker, T. R., 1974. Formation of red beds in most tropical climates: A hypothesis. *Geol. Soc. Amer. Bull.*, *85*, 633-648.

Walker, T. R., 1975. Intrastratal sources of uranium in first-cycle, nonmarine red beds (Abs). *Amer. Assoc. Petrol. Geol. Bull.*, *59*, 925.

Walker, G. W., Osterwald, F. W., and Adams, J. W., 1963. Geology of uranium-bearing veins in the conterminous United States. *U. S. Geol. Surv. Prof. Ppr.* *455*, 146 pp.

Warren, C. G., 1972. Sulfur isotopes as a clue to the genetic geochemistry of a roll-type uranium deposit. *Econ. Geol.*, *67*, 759-767.

Waters, A. C., and Granger, H. C., 1953. Volcanic debris in uraniferous sandstones and its possible bearing on the origin and precipitation of uranium. *U. S. Geol. Surv. Circ.* *224*.

Weber, W., and Stephenson, J. F., 1973. The content of mercury and gold in some Archean rocks of Rice Lake (Manitoba) area. *Econ. Geol.*, *68*, 401-407.

Wedepohl, K. H., 1964. Untersuchen am Kupferschiefer in NW Deutschland: Ein Beitrag zur Deutung der Genese bituminöser Sedimente. *Geochim. Cosmochim. Acta*, *28*, 305-309.

Weeks, A. D., 1956. Mineralogy and oxidation of the Colorado Plateau uranium ores. *U. S. Geol. Survey Prof. Ppr.* *300*, 187-193.

Weeks, A. D., and Thompson, M. E., 1954. Identification and occurrence of uranium and vanadium minerals from the Colorado Plateau. *U. S. Geol. Surv. Bull.* *1009-B*.

White, L., 1975a. Wyoming uranium miners set sights on higher production. *Eng. Min. J.*, *176(12)*, 61-71.

White, L., 1975b. In-situ leaching opens new uranium reserves in Texas. Eng. Min. J., 176(7), 73-81.

Wilshire, H. G., and Pike, J. E. N., 1975. Upper-mantle diapirism: Evidence from analogous features in alpine peridotite and ultramafic inclusions in basalt. Geology, 3, 467-470.

Winterer, E. L., 1973. Sedimentary facies and plate tectonics of equatorial Pacific. Amer. Assoc. Petrol. Geol. Bull., 57, 265-282.

Wright, R. J., 1955. Ore controls in sandstone uranium deposits of the Colorado Plateau. Econ. Geol., 50, 135-155.

Wright, J. B., and McCurry, P., 1973. Sea-floor spreading and continental ore deposits. In: Tarling and Runcorn (Editors) (q. v.), vol. 1, pp. 563-565.

Wyllie, P. J. (Editor), 1967. Ultramafic and Related Rocks. Wiley, New York, 464 pp.

Wyllie, P. J., 1969a. The origin of ultramafic and ultrabasic rocks. Tectonophys., 7, 437-455.

Wyllie, P. J., 1969b. The ultramafic belts. In: Hart (Editor) (q. v.), pp. 480-488.

Wyllie, P. J., 1971. The Dynamic Earth: Textbook in Geosciences. Wiley, New York, 416 pp.

Wynne-Edwards, H. R. (Editor), 1969. Age Relations in High-Grade Metamorphic Terrains. Geol. Soc. Can., Toronto, 288 pp.

Yanshin, A. L., 1975. Composition of Phosphate Crusts of Weathering and Associated Phosphate Deposits. USSR Acad. Sci., Novosibirsk, 210 pp.

York, J. E., and Helmberger, D. V., 1973. Low-veolocty zone variations in the southwestern United States. J. Geophys. Res., 78, 1883-1886.

Yoshii, T., 1975. Regionality of group velocities of Rayleigh waves in the Pacific and thickening of the plate. Earth Planet. Sci. Ltr., 25, 305-312.

Young, R. G., 1964. Distribution of uranium deposits in White Canyon - Monument Valley district, Utah-Arizona. Econ. Geol., 59, 850-873.

Zapletal, K., 1968. Tektogene der Erde. In: M. Mistik (Editor), Orogenic Belts. Academia, Prague, pp. 213-224.

Ziegler, V., 1974. Essai de classification metallotectonique des gisements d'uranium. In: IAEA (q. v.), pp. 661-677.

Zonenshain, L. P., 1971. Geosynclinal processes and the new global tectonics. Geotectonics, 6, 3-26.

Zonenshain, L. P., 1972. Similarities in the evolution of geosynclines of different types. Internat. Geol. Cong. Abstr., 24, 106.

Zonenshain, L. P., 1973. The evolution of central Asiatic geosynclines through sea-floor spreading. Tectonophys., 19, 213-232.

Zonenshain, L. P., Kuzmin, M. I., Kovalenko, V. I., and Saltykovsky, A. J., 1974. Mesozoic structural-magmatic pattern and metallogeny of the western part of the Pacific belt. Earth Planet. Sci. Ltr., 22, 96-109.

INDEX